Lecture Notes in Computer Science 12300

More information about this series at http://www.springer.com/series/7407

Vivek Nigam · Tajana Ban Kirigin ·
Carolyn Talcott · Joshua Guttman ·
Stepan Kuznetsov · Boon Thau Loo ·
Mitsuhiro Okada (Eds.)

Logic, Language, and Security

Essays Dedicated to Andre Scedrov
on the Occasion of His 65th Birthday

 Springer

Editors
Vivek Nigam (iD)
fortiss GmbH
Munich, Germany

Tajana Ban Kirigin (iD)
University of Rijeka
Rijeka, Croatia

Carolyn Talcott (iD)
SRI International
Menlo Park, CA, USA

Joshua Guttman
Worcester Polytechnic Institute
Worcester, MA, USA

Stepan Kuznetsov (iD)
Steklov Mathematical Institute
of the Russian Academy of Sciences
Moscow, Russia

Boon Thau Loo
University of Pennsylvania
Philadelphia, PA, USA

Mitsuhiro Okada (iD)
Keio University
Tokyo, Japan

ISSN 0302-9743   ISSN 1611-3349   (electronic)
Lecture Notes in Computer Science
ISBN 978-3-030-62076-9   ISBN 978-3-030-62077-6   (eBook)
https://doi.org/10.1007/978-3-030-62077-6

LNCS Sublibrary: SL1 – Theoretical Computer Science and General Issues

Photo by Nikola Ščedrov

Preface

It is an honor to be associated with Andre Scedrov as a scientific collaborator and a friend. While numbers do not completely reflect Andre's ability in building bridges between researchers and between different topics, they are nevertheless impressive. With an Erdös Number of 2, Andre Scedrov has co-authored papers with more than 90 researchers, distributed among his more than 130 refereed publications in the last 40 years of contributions. His research has impacted scientists from the four corners of the world, including the USA, Brazil, Japan, Russia, and many countries in Europe, including Croatia, his birth place.

Scedrov helped establish important venues, such as the IEEE Symposium on Logic in Computer Science (LICS), being actively part of its Advisory Board (1998–2012), and the Computer Security Foundations Workshop (CSFW) and its derived conference IEEE Computer Security Foundations Symposium (CSF).

He has been a member of the Mathematics Department at the University of Pennsylvania, USA, since the 1980s, and is currently a full professor, having recently served as chair. He is also a professor at the Department of Computer and Information Science.

Scedrov's contributions are particularly substantial, since his work transformed three very separate fields, namely linear logic and structural proof theory; formal reasoning for networked systems; and foundations of information security emphasizing cryptographic protocols.

Scedrov carried out seminal work in the early days of linear logic and structural proof theory that have shaped linear logic research for the following decades. Together with Jean-Yves Girard and Philip Scott, he developed Bounded Linear Logic, establishing foundational connections between linear logic and the complexity of algorithms. This research has evolved into a field of its own called implicit computational complexity where his initial ideas can still be observed. Scedrov also had important contributions to the influential program of studying the complexity of the provability problem of fragments of linear logic. This seminal work greatly advanced our understanding of computational limits of linear logic theories and still impacts research on the field. Finally, Scedrov contributed to the foundations of logic programming, helping understand logic programming from a proof-theoretic perspective. This work has set the stage for a number of research programs on computational logic and logical frameworks.

Scedrov published a series of papers that aim to bridge the gap between formal reasoning and networked systems. Back in the early 2010s, when network verification was a nascent field, together with his student Anduo Wang and colleagues at Penn, Carnegie-Mellon University, and SRI, he developed Formally Safe Routing, a toolkit that can generate verified Internet routing protocol implementations using a combination of declarative networking techniques, in conjunction with Yices SMT solvers, PVS theorem provers, and the Maude rewriting engine. This line of work is significant as it aims to bridge the gap between formal methods and network implementations. Today, there are entire tracks at major networking conferences (NSDI, SIGCOMM) on network verification, particularly in the context of software-defined networks.

In the foundations of security, Scedrov has worked on three major lines of work. First, in collaboration with John Mitchell, Iliano Cervesato, Joe-Kai Tsay, and others, he developed the use of multiset rewriting as a formalism for reasoning about security protocols. This led to major theoretical conclusions about the complexity of reasoning about protocols, as well as major applications to widely deployed protocols including Kerberos. He and his group saved the IETF from deploying a flawed public-key initiation protocol for Kerberos by discovering its failure before standardization had completed. More recently, multiset rewriting has been very well implemented in a protocol analysis tool developed at ETH Zurich, Switzerland. In a separate line of work, he explored the boundaries between the formal protocol analysis we have been discussing so far and a harder but more high-fidelity method. This is the computational foundation of cryptography that cryptographers use. When applicable, the latter gives strong guarantees; however, the formal analysis is far easier to use or even mechanize. Working with his PhD students Gergei Bana and Pedro Adão, Scedrov developed strong methods for showing that the simpler formal technique would yield the same conclusions as the more arduous computational technique. The third line of work – in collaboration with his student Rohit Chadha and also Steve Kremer – developed techniques for reasoning about contract-signing protocols, which have subtler goals than key agreement security protocols.

This combination of breadth and penetrating originality is rare and impressive. This Festschrift in his honor only tries to reflect this combination with a number of contributions distributed among these different topics.

Finally, we are in great debt to the authors for their excellent submissions, and the hard work done by the paper reviewers in providing reviews within our tight schedules.

July 2020

Vivek Nigam
Tajana Ban Kirigin
Carolyn Talcott
Joshua Guttman
Stepan Kuznetsov
Boon Thau Loo
Mitsuhiro Okada

Contents

Logic and Language

Logic

A Π^0_1-Bounded Fragment of Infinitary Action Logic with Exponential

Stepan L. Kuznetsov[1,2(✉)]

[1] Steklov Mathematical Institute of RAS, Moscow, Russia
sk@mi-ras.ru
[2] National Research University Higher School of Economics, Moscow, Russia

Abstract. Infinitary action logic is an extension of the multiplicative-additive Lambek calculus with Kleene iteration, axiomatized by an ω-rule. Buszkowski and Palka (2007) show that this logic is Π^0_1-complete. As shown recently by Kuznetsov and Speranski, the extension of infinitary action logic with the exponential modality is much harder: Π^1_1-complete. The raise of complexity is of course due to the contraction rule. We investigate fragments of infinitary action logic with exponential, which still include contraction, but have lower (e.g., arithmetically bounded) complexity. In this paper, we show an upper Π^0_1 bound for the fragment of infinitary action logic, in which the exponential can be applied only to formulae of implication depth 0 or 1.

Keywords: Infinitary action logic · Exponential modality · Complexity · Lambek calculus

This paper is dedicated to Andre Scedrov on the occasion of his 65th birthday, with many thanks for fruitful collaboration and his exceptional generosity, as suggested by his last name: in Russian, 'generosity' is 'ščedrost'.

1 Introduction

The notion of *action lattice,* or residuated Kleene lattice, was introduced by Pratt [23] and Kozen [15]. Action lattices combine algebraic structures of residuated lattices (see [7]), with the notion of residuals going back to Krull [17], and Kleene algebras, the idea of which goes back to Kleene [14].

The inequational theory of action lattices is the set of all generally true inequations of the form $u \preceq v$, where u and v are terms in the language of action lattices. This theory is also called *action logic,* as it can be viewed as, and axiomatized as, a substructural propositional logic. This logic is an extension of the multiplicative-additive Lambek calculus [10,20] with the Kleene star as a unary connective. The Lambek calculus, in its turn, can be considered as an intuitionistic non-commutative version of Girard's [8] linear logic (see [1]). Linear logic also includes the exponential modality, which enables structural rules

© Springer Nature Switzerland AG 2020
V. Nigam et al. (Eds.): Scedrov Festschrift, LNCS 12300, pp. 3–16, 2020.
https://doi.org/10.1007/978-3-030-62077-6_1

(permutation, contraction, weakening), thus it is natural to extend the Lambek calculus with exponential [9,10,21] or even a family of subexponentials [12].

In action logic, Kleene star is axiomatized by induction-style axioms (for example, Pratt formulated them as 'pure induction,' $(p / p)^* \preceq p / p$). We consider a stronger system, called *infinitary action logic* [3,22], where Kleene star is axiomatized by an ω-rule. The motivation for this is as follows. Action logic itself is already Σ_1^0-complete [18], thus adding (sub)exponential modalities, which are axiomatized by finitary rules, could not increase complexity of the derivability problem. For infinitary action logic, the situation is different. Infinitary action logic without exponential is Π_1^0-complete [3,22]. However, if we enrich it with an unrestricted exponential modality, the derivability problem becomes much harder: Π_1^1-complete [19].

The main source of this complexity raising from Π_1^0 to Π_1^1 is, of course, the contraction rule for exponential. Contraction allows duplicating formulae and, thus, allows encoding of derivations from extra axioms by a variant of deduction theorem (see [19]). The problem of derivability from finite sets of extra axioms (derivability of Horn clauses) is, in its turn, Π_1^1-complete even for Kleene algebras without division operations, as shown by Kozen [16].

In this paper, we present a restricted fragment of infinitary action logic with exponential, where the exponential modality can be applied only to formulae of implication (division) depth less or equal to 1. The extension of the Lambek calculus with such a restricted exponential was studied by Fofanova and shown to be decidable [5,6] (for a previous result for depth 0 see [11]). We shall prove that the corresponding extension of infinitary action logic belongs to Π_1^0. This result is in the same line as Fofanova's one. Namely, though contraction is allowed, an exponential restricted to formulae of depth 0 or 1 does not lead to complexity growth.

2   Infinitary Action Logic with Exponential

Let us formulate infinitary action logic with exponential, denoted by $!\mathrm{ACT}_\omega$, as an infinitary sequent calculus. The system $!\mathrm{ACT}_\omega$, as defined below, is a fragment of a richer system from [19], which includes a family of subexponential modalities.

Formulae of $!\mathrm{ACT}_\omega$ are built from variables $(p, q, r, \ldots)$ and the unit constant $\mathbf{1}$ using five binary connectives: $\cdot$ (product), $\backslash$ (left division), $/$ (right division), $\vee$ (additive disjunction), and $\wedge$ (additive conjunction), and two unary connectives: $*$ (Kleene star) and $!$ (exponential). Sequents are expressions of the form $\Gamma \vdash C$, where C (succedent) is a formula and Γ (antecedent) is a sequence of formulae. The antecedent could be empty; in this case we write just $\vdash C$.

Axioms and inference rules of $!\mathrm{ACT}_\omega$ are as follows:

$$\frac{}{p \vdash p} \; Id$$

$$\frac{\Pi \vdash A \quad \Gamma, B, \Delta \vdash C}{\Gamma, \Pi, A \backslash B, \Delta \vdash C} \; \backslash L \qquad \frac{A, \Pi \vdash B}{\Pi \vdash A \backslash B} \; \backslash R$$

$$\frac{\Pi \vdash A \quad \Gamma, B, \Delta \vdash C}{\Gamma, B \,/\, A, \Pi, \Delta \vdash C} \;/L \qquad \frac{\Pi, A \vdash B}{\Pi \vdash B \,/\, A} \;/R$$

$$\frac{\Gamma, A, B, \Delta \vdash C}{\Gamma, A \cdot B, \Delta \vdash C} \;\cdot L \qquad \frac{\Gamma \vdash A \quad \Delta \vdash B}{\Gamma, \Delta \vdash A \cdot B} \;\cdot R$$

$$\frac{\Gamma, \Delta \vdash C}{\Gamma, 1, \Delta \vdash C} \;1L \qquad \frac{}{\vdash 1} \;1R$$

$$\frac{\Gamma, A_1, \Delta \vdash C \quad \Gamma, A_2, \Delta \vdash C}{\Gamma, A_1 \vee A_2, \Delta \vdash C} \;\vee L \qquad \frac{\Pi \vdash A_i}{\Pi \vdash A_1 \vee A_2} \;\vee R, \, i = 1, 2$$

$$\frac{\Gamma, A_i, \Delta \vdash C}{\Gamma, A_1 \wedge A_2, \Delta \vdash C} \;\wedge L, \, i = 1, 2 \qquad \frac{\Pi \vdash A_1 \quad \Pi \vdash A_2}{\Pi \vdash A_1 \wedge A_2} \;\wedge R$$

$$\frac{(\Gamma, A^n, \Delta \vdash C)_{n \in \mathbb{N}}}{\Gamma, A^*, \Delta \vdash C} \;*L \qquad \frac{\Pi_1 \vdash A \quad \ldots \quad \Pi_n \vdash A}{\Pi_1, \ldots, \Pi_n \vdash A^*} \;*R, \, n \geq 0$$

$$\frac{\Gamma, A, \Delta \vdash C}{\Gamma, !A, \Delta \vdash C} \;!L \qquad \frac{!A_1, \ldots, !A_n \vdash B}{!A_1, \ldots, !A_n \vdash \, !B} \;!R$$

$$\frac{\Gamma, \Pi, !A, \Delta \vdash C}{\Gamma, !A, \Pi, \Delta \vdash C} \;!P_1 \qquad \frac{\Gamma, !A, \Pi, \Delta \vdash C}{\Gamma, \Pi, !A, \Delta \vdash C} \;!P_2$$

$$\frac{\Gamma, \Delta \vdash C}{\Gamma, !A, \Delta \vdash C} \;!W \qquad \frac{\Gamma, !A, !A, \Delta \vdash C}{\Gamma, !A, \Delta \vdash C} \;!C$$

$$\frac{\Pi \vdash A \quad \Gamma, A, \Delta \vdash C}{\Gamma, \Pi, \Delta \vdash C} \;Cut$$

The *Id* axiom here is formulated in an atomic form, for convenience of derivation analysis. The general form of this axiom, $A \vdash A$ for an arbitrary A, is derivable. This is performed by induction on the structure of A. The interesting cases are ! and *. For !, the sequent $!A \vdash !A$ is derived from $A \vdash A$ by $!L$ and $!R$. For *, wishing to establish $A^* \vdash A^*$, we first derive all sequents of the form $A^n \vdash A^*$ (using *R), and then apply *L.

The cut rule is eliminable, as shown in [19] by combining methods of [22] and [12]. The system $!ACT_\omega$, in its full power, is Π^1_1-complete [19]. On the other hand, its fragment without ! (the basic infinitary action logic ACT_ω) belongs to Π^0_1, as shown by Palka [22]. We strengthen Palka's result by allowing exponential, restricted to a class of formulae.

Under !, we consider formulae constructed using only $\backslash$ and $/$ of implication (division) depth 0 or 1. These are formulae of the following form:

$$A = r_1 \backslash r_2 \backslash \ldots \backslash r_k \backslash p / q_1 / q_2 / \ldots / q_\ell.$$

(Here $\backslash$ associates to the right and $/$ associates to the left.) Here k and/or ℓ could be zero (if $k = \ell = 0$, we get just a variable, p).

Since the exponential modality allows permutation, $!(E \backslash F)$ is equivalent to $!(F / E)$. Thus, $!A$ is equivalent to $!(p / q_1 / q_2 / \ldots / q_\ell / r_1 / r_2 / \ldots / r_k)$, and we can consider, under !, only formulae with $k = 0$, of the form $p / q_1 / \ldots / q_n$ (here $n = k + \ell$ and $q_{\ell+i} = r_i$). We shall write such a formula as follows: $p / q_n \ldots q_1$, which reflects the fact that it is equivalent to $p / (q_n \cdot \ldots \cdot q_1)$.

Theorem 1. *The derivability problem in* $!\mathrm{ACT}_\omega$ *for sequents in which* ! *is applied only to formulae of the form* $p / q_1 \ldots q_n$ *belongs to the* Π_1^0 *class.*

Notice that, due to subformula property, the restriction on formulae under ! also holds for all sequents in a cut-free derivation of a restricted sequent.

By conservativity, Buszkowski's Π_1^0 lower bound for ACT_ω [3] also propagates to the restricted fragment of $!\mathrm{ACT}_\omega$. Thus, we conclude that the derivability problem in $!\mathrm{ACT}_\omega$ for sequents in which ! is applied only to formulae of the form $p / q_1 \ldots q_n$ is Π_1^0-complete.

The technique we use for proving Theorem 1 is a combination of two methods. The first one is Fofanova's method for proving decidability of the Lambek calculus extended with ! which can be applied only to formulae of implication depth 0 or 1 [5,6]. The second one is the *-elimination technique developed by Palka [22] for ACT_ω.

3   The Calculus $!_1^{\mathrm{D}}\mathrm{ACT}_\omega$

We start with reformulating the restricted fragment of $!\mathrm{ACT}_\omega$, following the ideas of [5,6]. The new system will be called $!_1^{\mathrm{D}}\mathrm{ACT}_\omega$. Here '1' means "implication depth ≤ 1" and 'D' stands for 'dyadic,' because in this calculus we use the idea of isolating !-formulae as in Andreoli's dyadic system Σ_2 [2].

Sequents of the dyadic system $!_1^{\mathrm{D}}\mathrm{ACT}_\omega$ have two zones in antecedents, for !-formulae and for normal (non-commutative linear) formulae. Thus, sequents are expressions of the form $\Phi; \Gamma \vdash C$, where C is a formula, Γ is a sequence of formulae, and Φ is a *set* of formulae. All formulae in Φ are required to be of the restricted form $p / q_1 \ldots q_n$. If Γ is empty, we write $\Phi; \vdash C$. A sequent with an empty Φ is written as $\Gamma \vdash C$ (which is a shortcut for $\varnothing; \Gamma \vdash C$). A sequent with a completely empty antecedent (both zones empty) is written as $\vdash C$.

If compared with the focused system for the multiplicative-additive Lambek calculus with subexponentials [13], our $!_1^{\mathrm{D}}\mathrm{ACT}_\omega$ does not maintain focus on non-!-formulae (thus, it is not a focused system). However, for !-formulae it behaves quite aggressively. Namely, !-formulae are confined in !-zones, and their decomposition is postponed up to axioms.

There are no explicit structural rules for ! in $!_1^D\mathrm{ACT}_\omega$. Permutability of !-formulae is now maintained by the structure of sequents: being a set, Φ is commutative by design. Weakening and, surprisingly, *contraction* is implemented in *axioms*.

Axioms are now quite complicated (not just $p \vdash p$, as in $!\mathrm{ACT}_\omega$). For a set Φ of formulae of the form $p \,/\, q_1 \ldots q_n$ consider a context-free grammar $\mathcal{G}_\Phi$ with the following production rules:

$$p \Rightarrow q_1 \ldots q_n \qquad \text{for each formula } (p \,/\, q_1 \ldots q_n) \in \Phi.$$

In $\mathcal{G}_\Phi$, we neither distinguish terminal and non-terminal symbols (the united alphabet is the set of variables), nor designate a starting symbol. Now we declare as axioms all sequents of the form

$$\Phi; r_1, \ldots, r_m \vdash s,$$

where $r_1 \ldots r_m$ is derivable from s in $\mathcal{G}_\Phi$ (that is, $s \Rightarrow_G^* r_1 \ldots r_m$). Additionally, sequents

$$\Phi; \vdash \mathbf{1}$$

are also axioms, for any Φ. The set of axioms is decidable, using standard parsing algorithms for context-free grammars.

Rules for Lambek connectives ($\cdot, \backslash, /$), additives ($\vee, \wedge$), and Kleene star in $!_1^D\mathrm{ACT}_\omega$ are basically copied from $!\mathrm{ACT}_\omega$. The !-zone (Φ) is non-deterministically distributed between branches in $/L$, $\backslash L$, $\cdot R$, and $*R$, with implicit contractions applied. (In order to reduce proof search, one could just force Φ to be the same; here we do not need it.)

$$\frac{\Psi; \Pi \vdash A \quad \Phi; \Gamma, B, \Delta \vdash C}{\Phi \cup \Psi; \Gamma, \Pi, A \backslash B, \Delta \vdash C} \; \backslash L \qquad \frac{\Phi; A, \Pi \vdash B}{\Phi; \Pi \vdash A \backslash B} \; \backslash R$$

$$\frac{\Psi; \Pi \vdash A \quad \Phi; \Gamma, B, \Delta \vdash C}{\Phi \cup \Psi; \Gamma, B \,/\, A, \Pi, \Delta \vdash C} \; / L \qquad \frac{\Phi; \Pi, A \vdash B}{\Phi; \Pi \vdash B \,/\, A} \; / R$$

$$\frac{\Phi; \Gamma, A, B, \Delta \vdash C}{\Phi; \Gamma, A \cdot B, \Delta \vdash C} \; \cdot L \qquad \frac{\Phi; \Gamma \vdash A \quad \Psi; \Delta \vdash B}{\Phi \cup \Psi; \Gamma, \Delta \vdash A \cdot B} \; \cdot R$$

$$\frac{\Phi; \Gamma, \Delta \vdash C}{\Phi; \Gamma, \mathbf{1}, \Delta \vdash C} \; 1L$$

$$\frac{\Phi; \Gamma, A_1, \Delta \vdash C \quad \Phi; \Gamma, A_2, \Delta \vdash C}{\Phi; \Gamma, A_1 \vee A_2, \Delta \vdash C} \; \vee L \qquad \frac{\Phi; \Pi \vdash A_i}{\Phi; \Pi \vdash A_1 \vee A_2} \; \vee R, \, i = 1, 2$$

$$\frac{\Phi; \Gamma, A_i, \Delta \vdash C}{\Phi; \Gamma, A_1 \wedge A_2, \Delta \vdash C} \; \wedge L, \, i = 1, 2 \qquad \frac{\Phi; \Pi \vdash A_1 \quad \Pi \vdash A_2}{\Phi; \Pi \vdash A_1 \wedge A_2} \; \wedge R$$

$$\frac{(\varPhi; \varGamma, A^n, \varDelta \vdash C)_{n \in \mathbb{N}}}{\varPhi; \varGamma, A^*, \varDelta \vdash C} \;\; *L \qquad \frac{\varPhi_1; \varPi_1 \vdash A \;\; \ldots \;\; \varPhi_n; \varPi_n \vdash A}{\varPhi_1 \cup \ldots \cup \varPhi_n; \varPi_1, \ldots, \varPi_n \vdash A^*} \;\; *R, \; n \geq 0$$

The rules for ! are as follows:

$$\frac{\varPhi \cup \{A\}; \varGamma, \varDelta \vdash C}{\varPhi; \varGamma, !A, \varDelta \vdash C} \;\; !L \qquad \frac{\varPhi; \vdash B}{\varPhi; \vdash !B} \;\; !R$$

Recall that each formula under ! should be of the form $p \,/\, q_1 \ldots q_n$.

Finally, now we have two cut rules, one for each zone:

$$\frac{\varPsi; \varPi \vdash A \quad \varPhi; \varGamma, A, \varDelta \vdash C}{\varPhi \cup \varPsi; \varGamma, \varPi, \varDelta \vdash C} \;\; Cut \qquad \frac{\varPsi; \vdash A \quad \varPhi \cup \{A\}; \varDelta \vdash C}{\varPhi \cup \varPsi; \varDelta \vdash C} \;\; Cut!$$

In the next section we prove that both cut rules in $!^{\mathrm{D}}_1\mathrm{ACT}_\omega$ are eliminable and that $!^{\mathrm{D}}_1\mathrm{ACT}_\omega$ is equivalent to the restricted fragment of $!\mathrm{ACT}_\omega$. Further, in Sect. 5, we develop Palka's *-elimination technique for $!^{\mathrm{D}}_1\mathrm{ACT}_\omega$ and in this way prove that it belongs to $\varPi^0_1$. This finishes the proof of Theorem 1.

As suggested by the Reviewer, instead of introducing a sophisticated set of axioms for $!^{\mathrm{D}}_1\mathrm{ACT}_\omega$ one could equivalently add inference rules of the form

$$\frac{\varPhi; \varPi_1 \vdash q_1 \;\; \ldots \;\; \varPhi; \varPi_n \vdash q_n}{\varPhi; \varPi_1, \ldots, \varPi_n \vdash p} \quad \text{provided that } p \,/\, q_1 \ldots q_n \in \varPhi.$$

This type of inference rule is a generalization of the rule called Buszkowski's rule B_1 in [11]. Buszkowski's rules are friendly to cut elimination [11]. However, they are not that convenient for inductive arguments on derivations: namely, if $n = 1$, then the rule does not reduce complexity when looking from bottom to top. Moreover, if $\varPhi$ includes both $p \,/\, q$ and $q \,/\, p$, then a derivation can include a useless sequence of rule applications replacing p with q and then back q with p. This would make our complexity upper bound argument in Sect. 5 problematic. In our formulation, all these issues are hidden into the context-free derivations used for verifying axioms, and we refer to ready-made context-free parsing algorithms. (In particular, issues with $p \,/\, q$ and $q \,/\, p$ are resolved by the standard chain rule elimination technique.)

4 Cut Elimination in $!^{\mathrm{D}}_1\mathbf{ACT}_\omega$ and Equivalence with Restricted $!\mathbf{ACT}_\omega$

Let us first establish cut elimination for $!^{\mathrm{D}}_1\mathrm{ACT}_\omega$.

Theorem 2. *Any sequent provable in* $!^{\mathrm{D}}_1\mathrm{ACT}_\omega$ *can be proved without using Cut and Cut!.*

Proof. As usual in cut elimination proofs, we shall proceed by induction on (cut-free) derivations of the premises of cut. Due to the infinitary nature of $!^{\mathrm{D}}_1\mathrm{ACT}_\omega$, these inductive arguments ought to be transfinite. For a cut-free derivation, we define its *rank* as an ordinal in the following way:

- the rank of an axiom is zero;
- if the lowermost rule is a finitary one (that is, any rule except $*L$), then the rank is the maximum of the ranks of the derivations of its premises, plus one;
- if the lowermost rule is $*L$, then the rank is the supremum of the ranks of the derivations of its (ω many) premises, plus one.

Thus, the rank reduces (at least by one) when moving from a derivation tree to a subtree deriving one of the premises of the lowermost rule.

We start with eliminating $Cut!$, which can be done independently from Cut. Recall that $A = p \, / \, q_1 \ldots q_n$. The left premise of $Cut!$, enjoys a cut-free derivation. Notice that no rule introduces formulae into the !-zone Ψ (this is done only in axioms). Thus, $\Psi; \vdash p \, / \, q_1 \ldots q_n$ was derived by n applications of $/R$ from the axiom $\Psi; q_1 \ldots q_n \vdash p$.

Now proceed by induction on the cut-free derivation of $\Phi \cup \{A\}; \Delta \vdash C$. The key consideration here is that all rule applications are non-principal and can be exchanged with $Cut!$ (this reduces the rank). The reason is that the cut formula A is in the !-zone and therefore is kept intact up to axioms.

Finally, let $\Phi \cup \{A\}; \Delta \vdash C$ be an axiom of the form $\Phi \cup \{A\}; r_1, \ldots, r_m \vdash s$. Recall that $\Psi; q_1, \ldots, q_n \vdash p$ is also an axiom, and therefore $p \Rightarrow_{\mathcal{G}_\Psi} q_1 \ldots q_n$. This is exactly the context-free production rule expressed by formula A. Given $s \Rightarrow^*_{\mathcal{G}_{\Phi \cup \{A\}}} r_1 \ldots r_m$, let us replace each application of $p \Rightarrow q_1 \ldots q_n$ with the corresponding derivation in $\mathcal{G}_\Psi$. This yields $s \Rightarrow^*_{\Phi \cup \Psi} r_1 \ldots r_m$. Therefore, the sequent $\Phi \cup \Psi; r_1, \ldots, r_m \vdash s$, which is the goal of $Cut!$, is an axiom.

Now we eliminate Cut, proceeding by a standard induction (but in a transfinite setting). The two parameters are now α and β, the ranks of cut-free derivations of the premises of Cut. Notice that the following preorder on pairs (α, β) is well-founded: $(\alpha_1, \beta_1) \prec (\alpha_2, \beta_2)$ if and only if either $\alpha_1 < \alpha_2$ and $\beta_1 \leq \beta_2$, or $\alpha_1 \leq \alpha_2$ and $\beta_1 < \beta_2$. At each step (α, β) will be $\prec$-reduced.

In non-principal cases, where at least one of the lowermost rules in derivations of $\Psi; \Pi \vdash A$ and $\Phi; \Gamma, A, \Delta \vdash C$ does not introduce A, cut gets propagated, lowering one of the ranks while keeping the other one. Principal cases for all connectives, except !, are considered as in Palka's proof [22], which generally follows Lambek's scheme [20].

Finally, for ! the principal case is as follows:

$$\dfrac{\dfrac{\Psi; \vdash A}{\Psi; \vdash !A} \, !R \qquad \dfrac{\Phi \cup \{A\}; \Gamma, \Delta \vdash C}{\Phi; \Gamma, !A, \Delta \vdash C} \, !L}{\Phi \cup \Psi; \Gamma, \Delta \vdash C} \, Cut$$

and Cut is transformed into $Cut!$:

$$\dfrac{\Psi; \vdash A \qquad \Phi \cup \{A\}; \Gamma, \Delta \vdash C}{\Phi \cup \Psi; \Gamma, \Delta \vdash C} \, Cut!$$

Premises of $Cut!$ have cut-free proofs, and we can eliminate it as shown above.

This shows how to eliminate one (lowermost) instance of cut. In order to eliminate many (maybe infinitely many) cuts in a derivation, we define the notion of *cut rank*. Cut rank is defined similarly to the rank of a derivation, but now we count only cuts. That is, for rules except *Cut* and *Cut*! the cut rank is just counted as the supremum (maximum) of the cut ranks for premises, and for *Cut* and *Cut*! this maximum is increased by one.

Proofs in $!_1^D\mathrm{ACT}_\omega$ do not have infinite branches. Thus, we have a set of *topmost cuts*, whose premises enjoy cut-free derivations. Eliminating these cuts reduces the cut rank. □

Now we show equivalence between $!_1^D\mathrm{ACT}_\omega$ and the restricted fragment of $!\mathrm{ACT}_\omega$. We start with a technical lemma.

Lemma 1. *Let* $A = p \mathbin{/} q_1 \ldots q_n$. *Then the sequents* $\{A\}; \vdash A$ *and* $\{A\}; \vdash \,!A$ *are derivable in* $!_1^D\mathrm{ACT}_\omega$.

Proof. The sequent $\{A\}; \vdash A$ is derived from $\{A\}; q_1, \ldots, q_n \vdash p$ by n applications of $/R$. The latter sequent is an axiom, since $q_1 \ldots q_n$ is derived from p in $\mathcal{G}_{\{A\}}$ in one step. The second sequent, $\{A\}; \vdash \,!A$, is derived from $\{A\}; \vdash A$ by applying $!R$. □

Theorem 3. *Let* $\Gamma \vdash C$ *be a sequent in which* ! *is applied only to formulae of the form* $p \mathbin{/} q_1 \ldots q_n$. *Then* $\Gamma \vdash C$ *is derivable in* $!\mathrm{ACT}_\omega$ *if and only if* $\Gamma \vdash C$ *is derivable in* $!_1^D\mathrm{ACT}_\omega$.

Proof. For the "only if" part, notice that the rules of $!\mathrm{ACT}_\omega$ not operating ! directly map onto the corresponding rules of $!_1^D\mathrm{ACT}_\omega$ with empty !-zones added. The remaining rules $!L$, $!R$, $!P_1$, $!P_2$, $!W$, and $!C$ are modelled using cut and Lemma 1 as follows:

$$\cfrac{\cfrac{\{A\}; \vdash A \quad \Gamma, A, \Delta \vdash C}{\{A\}; \Gamma, \Delta \vdash C}\ Cut}{\Gamma, !A, \Delta \vdash C}\ !L$$

$$\cfrac{\cfrac{\cfrac{(\{A_i\}; \vdash \,!A_i)_{i=1}^n \quad !A_1, \ldots, !A_n \vdash B}{\{A_1, \ldots, A_n\}; \vdash B}\ Cut\ (n\ \text{times})}{\{A_1, \ldots, A_n\}; \vdash \,!B}\ !R}{!A_1, \ldots, !A_n \vdash \,!B}\ !L\ (n\ \text{times})$$

$$\cfrac{\cfrac{\{A\}; \vdash \,!A \quad \Gamma, \Pi, !A, \Delta \vdash C}{\{A\}; \Gamma, \Pi, \Delta \vdash C}\ Cut}{\Gamma, !A, \Pi, \Delta \vdash C}\ !L \qquad \cfrac{\cfrac{\{A\}; \vdash \,!A \quad \Gamma, !A, \Pi, \Delta \vdash C}{\{A\}; \Gamma, \Pi, \Delta \vdash C}\ Cut}{\Gamma, !A, \Pi, \Delta \vdash C}\ !L$$

$$\cfrac{\cfrac{\{A\}; \vdash 1 \quad \cfrac{\Gamma, \Delta \vdash C}{\Gamma, 1, \Delta \vdash C}\ 1L}{\{A\}; \Gamma, \Delta \vdash C}\ Cut}{\Gamma, !A, \Delta \vdash C}\ !L$$

$$\cfrac{\{A\};\ \vdash\ !A \qquad \cfrac{\{A\};\ \vdash\ !A \qquad \Gamma, !A, !A, \Delta \vdash C}{\{A\}; \Gamma, !A, \Delta \vdash C}\ Cut}{\cfrac{\{A\}; \Gamma, \Delta \vdash C}{\Gamma, !A, \Delta \vdash C}\ !L}\ Cut$$

For the "if" part, we prove a stronger statement: if $\{A_1, \ldots, A_n\}; \Gamma \vdash C$ is derivable in $!_1^{\mathrm{D}}\mathrm{ACT}_\omega$, then $!A_1, \ldots, !A_n, \Gamma \vdash C$ is derivable in $!\mathrm{ACT}_\omega$. This is proved by direct transformation of the cut-free derivation of $\{A_1, \ldots, A_n\}; \Gamma \vdash C$ in $!_1^{\mathrm{D}}\mathrm{ACT}_\omega$.

The most interesting case is the axiom. Recall that axioms of $!_1^{\mathrm{D}}\mathrm{ACT}_\omega$ are sequents of the form $\Phi; r_1, \ldots, r_m \vdash s$, where $r_1 \ldots r_m$ is derivable from s in the context-free grammar $\mathcal{G}_\Phi$. Let $\Phi = \{A_1, \ldots, A_n\}$ and proceed by induction on this context-free derivation (this is a regular induction, not a transfinite one). If $r_1 \ldots r_m = s$, then $!A_1, \ldots, !A_n, s \vdash s$ is derivable from axiom Id by n applications of $!W$. If the first production rule is $s \Rightarrow q_1 \ldots q_k$, then the following sequents are derivable by induction hypothesis:

$$!\Phi, r_1, \ldots, r_{j_1} \vdash q_1, \qquad !\Phi, r_{j_1+1}, \ldots, r_{j_2} \vdash q_2, \qquad \ldots, \qquad !\Phi, r_{j_{k-1}+1}, \ldots, r_m \vdash q_k,$$

where $!\Phi = !A_1, \ldots, !A_n$. Now $!\Phi, r_1, \ldots, r_m \vdash s$ is derived as follows (recall that $A = s \,/\, q_1 \ldots q_k$ belongs to Φ):

$$\cfrac{\cfrac{!\Phi, r_1, \ldots, r_{j_1} \vdash q_1 \quad \ldots \quad !\Phi, r_{j_{k-1}+1}, \ldots, r_m \vdash q_k \quad s \vdash s}{s \,/\, q_1 \ldots q_k, !\Phi, r_1, \ldots, r_{j_1}, \ldots, !\Phi, r_{j_{k-1}+1}, \ldots, r_m \vdash s}\ /\,L\ (k\ \text{times})}{\cfrac{!\Phi, s \,/\, q_1 \ldots q_k, r_1, \ldots, r_m \vdash s}{!\Phi, r_1, \ldots, r_m \vdash s}\ !C, !P, !L}\ !C, !P\ (\text{several times})$$

Axioms of the form $\Phi; \ \vdash 1$ are translated as $!A_1, \ldots, !A_n \vdash 1$ and derived from $\vdash 1$ by $!W$. Inference rules of $!_1^{\mathrm{D}}\mathrm{ACT}_\omega$ not operating $!$ are translated to the corresponding rules of $!\mathrm{ACT}_\omega$, adding $!P$ and $!C$, where necessary. Finally, $!L$ in $!_1^{\mathrm{D}}\mathrm{ACT}_\omega$ is a specific form of $!P$ in $!\mathrm{ACT}_\omega$, and $!R$ corresponds directly to $!R$. □

In the next section we show that the derivability problem for $!_1^{\mathrm{D}}\mathrm{ACT}_\omega$ belongs to Π_1^0. By the equivalence theorem, this yields the same complexity upper bound for the restricted fragment of $!\mathrm{ACT}_\omega$.

5 *-Elimination and Complexity of $!_1^{\mathrm{D}}\mathrm{ACT}_\omega$

In this section we develop Palka's [22] *-elimination technique for $!_1^{\mathrm{D}}\mathrm{ACT}_\omega$. Our exposition follows [4], rather than [22]. The crucial notion here is the notion of *n-th approximation* of a sequent. Informally, the n-th approximation is obtained by replacing each negative occurrence of B^* with $B^{\leq n} = 1 \vee B \vee B^2 \vee \ldots \vee B^n$

(where $B^n = \underbrace{B \cdot \ldots \cdot B}_{n\ \text{times}}$). The formal definition involves defining two functions, N_n and P_n, by joint induction[1]:

$$N_n(p) = P_n(p) = p \qquad\qquad N_n(1) = P_n(1) = 1$$
$$N_n(A \setminus B) = P_n(A) \setminus N_n(B) \qquad P_n(A \setminus B) = N_n(A) \setminus P_n(B)$$
$$N_n(B / A) = N_n(B) / P_n(A) \qquad P_n(B / A) = P_n(B) / N_n(A)$$
$$N_n(A \cdot B) = N_n(A) \cdot N_n(B) \qquad P_n(A \cdot B) = P_n(A) \cdot P_n(B)$$
$$N_n(A \vee B) = N_n(A) \vee N_n(B) \qquad P_n(A \vee B) = P_n(A) \vee P_n(B)$$
$$N_n(A \wedge B) = N_n(A) \wedge N_n(B) \qquad P_n(A \wedge B) = P_n(A) \wedge P_n(B)$$
$$N_n(B^*) = 1 \vee N_n(B) \vee (N_n(B))^2 \vee \ldots \vee (N_n(B))^n$$
$$N_n(!A) = P_n(!A) = !A \qquad\qquad P_n(B^*) = (P_n(B))^*$$

Notice that in $!_1^{\mathrm{D}}\mathrm{ACT}_\omega$ formulae of the form $!A$ never contain Kleene star, therefore we do not need to propagate N_n or P_n in this case.

The n-th approximation of the sequent of the form $\Phi; B_1, \ldots, B_k \vdash C$ is the sequent $\Phi; N_n(B_1), \ldots, N_n(B_k) \vdash P_n(C)$. The *-elimination theorem (which eliminates all negative occurrences of Kleene star) is formulated as follows:

Theorem 4. *A sequent is derivable in $!_1^{\mathrm{D}}\mathrm{ACT}_\omega$ if and only for any n its n-th approximation is derivable in $!_1^{\mathrm{D}}\mathrm{ACT}_\omega$.*

Following [4], we first establish two lemmas.

Lemma 2. *The sequents $A \vdash P_n(A)$ and $N_n(A) \vdash A$ are derivable in $!_1^{\mathrm{D}}\mathrm{ACT}_\omega$ for any formula A.*

Proof. By induction on the complexity of A, exactly as in [4]. (Notice that the case of $A = !A'$ is trivial, since $N_n(!A') = P_n(!A') = !A'$ by definition.)   □

Lemma 3. *If $m \le n$, then $N_m(A) \vdash N_n(A)$ and $P_n(A) \vdash P_m(A)$ are derivable in $!_1^{\mathrm{D}}\mathrm{ACT}_\omega$ for any formula A.*

Proof. Also similar to [4].   □

Proof (of Theorem 4).

The "only if" part follows from Lemma 2 using cut.

For the more interesting "if" part, we proceed by nested induction on two complexity parameters for the given sequent $\Phi; B_1, \ldots, B_k \vdash C$. The parameter of the inner induction is the usual complexity, counted as the total number of connectives in $B_1, \ldots, B_k$, and C (we do not count the complexity of Φ, since formulae in Φ are just propagated up to axioms, and are never decomposed).

The parameter of the outer induction is more complicated. We call it *star rank* and define as follows. The star rank σ of a formula or a sequent is a sequence of natural numbers, which is formally infinite, but, starting from some point, includes only zeroes.

On such sequences, we define two operations:

---
[1] In [4], the notations N and P are inverted.

– if $\xi = (x_0, x_1, x_2, \ldots)$ and $\eta = (y_0, y_1, y_2, \ldots)$, then $\xi \oplus \eta = (x_0 + y_0, x_1 + y_1, x_2 + y_2, \ldots)$;

– if $\xi = (x_0, x_1, x_2, \ldots)$, then $\xi \uparrow = (0, x_0, x_1, x_2, \ldots)$.

The order on such sequences is lexicographical. It is easy to see that it is well-founded. Thus, we can use induction.

The inductive definition of σ is as follows:

$$\sigma(p) = \sigma(\mathbf{1}) = \sigma(!A) = 0;$$
$$\sigma(A \cdot B) = \sigma(A \backslash B) = \sigma(B \,/\, A) = \sigma(A \vee B) = \sigma(A \wedge B) = \sigma(A) \oplus \sigma(B);$$
$$\sigma(B^*) = \sigma(B) \uparrow \oplus (1, 0, 0, 0, \ldots).$$

Informally, the d-th component of $\sigma(A)$ is the number of occurrences of $*$ of nesting depth d. For a sequent $\Phi; B_1, \ldots, B_n \vdash C$, its star rank is $\sigma(B_1) \oplus \ldots \oplus \sigma(B_n) \oplus \sigma(C)$.

Now consider a sequent $\Phi; B_1, \ldots, B_k \vdash C$ and suppose that, for any n, the sequent $\Phi; N_n(B_1), \ldots, N_n(B_k) \vdash P_n(C)$ is derivable in $!_1^{\mathrm{D}}\mathrm{ACT}_\omega$. Consider two cases:

Case 1: one of B_i is of the form E^*. Then $N_n(B_i) = (N_n(E))^{\leq n}$. By cut with $(N_n(E))^m \vdash (N_n(E))^{\leq n}$, for $m \leq n$, we get derivability of

$$\Phi; N_n(B_1), \ldots, (N_n(E))^m, \ldots, N_n(B_k) \vdash P_n(C),$$

provided $n \geq m$. Using Lemma 3, we get derivability of this sequent for arbitrary n and m.

This new sequent, for a fixed m, has less star rank than the original sequent. Thus, by induction hypothesis we get derivability of

$$\Phi; B_1, \ldots, E^m, \ldots, B_k \vdash C,$$

and by the ω-rule $*L$ we derive our goal sequent $\Phi; B_1, \ldots, B_i, \ldots B_k \vdash C$ (recall that $B_i = E^*$).

Case 2: none of B_i is of the form E^*. Consider the lowermost rules in cut-free derivations of

$$\Phi; N_n(B_1), \ldots, N_n(B_k) \vdash P_n(C),$$

for $n = 0, 1, 2, \ldots$ If a least one of these sequents is an axiom, then it should coincide with $\Phi; B_1, \ldots, B_k \vdash C$, whence the latter is also an axiom. Otherwise, each of these sequents is derived using a one of the finitary rules (not $*L$).

The main connective of each $N_n(B_i)$ is the same as that of B_i itself (recall that $B_i \neq E^*$), ditto for $P_n(C)$. For each finitary rule, the complete information about its application is as follows: (1) the number i of B_i introduced by the rule, or $i = k + 1$ for the rule introducing C; (2) the lengths of the parts of the antecedent which go to its premises (for example, for $*R$ these are the lengths of $\Pi_1, \ldots, \Pi_n$); (3) the way Φ is distributed between premises. Let us call this collection of information the *form* of the rule.

For a given goal sequent $\Phi; B_1, \ldots, B_k \vdash C$, there is only a finite number of possible forms for lowermost rules in derivations of $\Phi; N_n(B_1), \ldots, N_n(B_k) \vdash C$. Therefore, there is an infinite set $\mathcal{N}$ of natural numbers, such that this sequent is derived using a rule of the same form for all $n \in \mathcal{N}$.

The premises of these rules can be defined uniformly, as n-th approximations of the corresponding sequents obtained from $\Phi; B_1, \ldots, B_k \vdash C$. Using Lemma 3, we prove these premises for *all* n, since for any n there exists $n' \in \mathcal{N}$ such that $n' \geq n$. Finally, we derive $\Phi; B_1, \ldots, B_k \vdash C$ using the rule of that form. The premises of this rule have either a smaller star rank, or, if the star rank is the same, smaller complexity. Thus, they are derivable by induction hypothesis. $\square$

Theorem 4 immediately yields Theorem 1. Indeed, n-th approximations include no negative occurrences of $*$, and derivability of such sequents is decidable (for checking axioms, we use standard context-free parsing algorithms); the outer "$\forall n$" quantifier yields Π_1^0.

Future Work

The use of Palka's $*$-elimination technique for $!_1^D \mathrm{ACT}_\omega$ became possible only because we managed to reformulate our calculus without an explicit contraction rule. In the presence of contraction, the $*$-elimination argument fails. Namely, in Case 2 of the proof of Theorem 4, if the rule we wish to apply is contraction, then its premise has greater complexity than the conclusion.

Actually, for $!\mathrm{ACT}_\omega$ in general $*$-elimination should not hold due to complexity reasons. Indeed, $*$-elimination would yield a Π_2^0 upper complexity bound: derivability in the fragment of $!\mathrm{ACT}_\omega$ without negative occurrences of $*$ is undecidable (cf. [21]), but belongs to Σ_1^0, and $*$-elimination would add just one "$\forall n$" quantifier. As shown in [19], however, $!\mathrm{ACT}_\omega$ is Π_1^1-complete, which is much more than Π_2^0.

In the view of the above, it is interesting to find a fragment of $!\mathrm{ACT}_\omega$, which is Π_2^0-complete. A natural candidate would be the fragment where $*$ is not allowed under $!$. For this system, a Π_2^0 lower bound is provided by $!$-encoding of derivability of sequents from finite sets of $*$-free extra axioms (finite $*$-free theories). The latter is Π_2^0-complete due to Kozen [16]. Reformulating this system for $*$-elimination to work, however, is an open problem.

A toy example of a Π_2^0-complete fragment, however, can be constructed as follows: sequents are required to be of the form $E_1, \ldots, E_k \vdash F$, where F is $*$-free and E_i is either $*$-free, or of the form G^* for a $*$-free G. Such sequents are still powerful enough to encode derivability in Kleene algebras from finite $*$-free theories (see [16]). This gives Π_2^0-hardness. On the other hand, all occurrences of $*$ are on the top level, and $*$-elimination comes trivially from invertibility of $*L$, thus the upper Π_2^0 boundary. Extending this to an interesting fragment is left for further research.

Financial Support. Research work towards this paper was supported by the HSE University Basic Research Program funded by the Russian Academic Excellence

Project '5-100,' by grant MK-430.2019.1 of the President of Russia, by the Young Russian Mathematics Award, and by the Russian Foundation for Basic Research grant 20-01-00435.

References

1. Abrusci, V.M.: A comparison between Lambek syntactic calculus and intuitionistic linear logic. Zeitschr. Math. Logik Grundl. Math. (Math. Log. Q.) **36**, 11–15 (1990)
2. Andreoli, J.-M.: Logic programming with focusing proofs in linear logic. J. Log. Comput. **2**(3), 297–347 (1992)
3. Buszkowski, W.: On action logic: equational theories of action algebras. J. Log. Comput. **17**(1), 199–217 (2007)
4. Buszkowski, W., Palka, E.: Infinitary action logic: complexity, models and grammars. Stud. Log. **89**(1), 1–18 (2008). https://doi.org/10.1007/s11225-008-9116-7
5. Fofanova, E.M.: Algorithmic decidability of a fragment of the Lambek calculus with exponential modality. In: Mal'tsev Meeting 2018, Collection of Abstracts, p. 226. Sobolev Institute of Mathematics and Novosibirsk State University, Novosibirsk (2018). (in Russian)
6. Fofanova, E.M.: Algorithmic decidability of a fragment of the Lambek calculus with exponential modality. M.Sc. thesis, Moscow State University (2019). (in Russian)
7. Galatos, N., Jipsen, P., Kowalski, T., Ono, H.: Residuated Lattices: An Algebraic Glimpse at Substructural Logics. Studies in Logic and the Foundations of Mathematics, vol. 151. Elsevier, Amsterdam (2007)
8. Girard, J.-Y.: Linear logic. Theor. Comput. Sci. **50**(1), 1–101 (1987)
9. de Groote, P.: On the expressive power of the Lambek calculus extended with a structural modality. In: Casadio, C., Scott, P.J., Seely, R.A.G. (eds.) Language and Grammar. Studies in Mathematical Linguisticsand Natural Language. CSLI Lecture Notes, vol. 168, pp. 95–111 (2005)
10. Kanazawa, M.: The Lambek calculus enriched with additional connectives. J. Log. Lang. Inf. **1**(2), 141–171 (1992). https://doi.org/10.1007/BF00171695
11. Kanovich, M., Kuznetsov, S., Scedrov, A.: Undecidability of the Lambek calculus with a relevant modality. In: Foret, A., Morrill, G., Muskens, R., Osswald, R., Pogodalla, S. (eds.) FG 2015-2016. LNCS, vol. 9804, pp. 240–256. Springer, Heidelberg (2016). https://doi.org/10.1007/978-3-662-53042-9_14
12. Kanovich, M., Kuznetsov, S., Nigam, V., Scedrov, A.: Subexponentials in non-commutative linear logic. Math. Struct. Comput. Sci. **29**(8), 1217–1249 (2019)
13. Kanovich, M., Kuznetsov, S., Nigam, V., Scedrov, A.: A logical framework with commutative and non-commutative subexponentials. In: Galmiche, D., Schulz, S., Sebastiani, R. (eds.) IJCAR 2018. LNCS (LNAI), vol. 10900, pp. 228–245. Springer, Cham (2018). https://doi.org/10.1007/978-3-319-94205-6_16
14. Kleene, S.C.: Representation of events in nerve nets and finite automata. In: Automata Studies, pp. 3–41. Princeton University Press (1956)
15. Kozen, D.: On action algebras. In: van Eijck, J., Visser, A. (eds.) Logic and Information Flow, pp. 78–88. MIT Press (1994)
16. Kozen, D.: On the complexity of reasoning in Kleene algebra. Inform. Comput. **179**(2), 152–162 (2002)
17. Krull, W.: Axiomatische Begründung der algemeinen Idealtheorie. Sitz. Phys.-Med. Soc. Erlangen **56**, 47–63 (1924)

18. Kuznetsov, S.: The logic of action lattices is undecidable. In: 34th Annual ACM/IEEE Symposium on Logic in Computer Science (LICS). IEEE (2019)
19. Kuznetsov, S.L., Speranski, S.O.: Infinitary action logic with exponentiation. arXiv preprint arXiv:2001.06863 (2020)
20. Lambek, J.: The mathematics of sentence structure. Am. Math. Mon. **65**, 154–170 (1958)
21. Lincoln, P., Mitchell, J., Scedrov, A., Shankar, N.: Decision problems for propositional linear logic. Ann. Pure Appl. Log. **56**(1–3), 239–311 (1992)
22. Palka, E.: An infinitary sequent system for the equational theory of *-continuous action lattices. Fundam. Inform. **78**(2), 295–309 (2007)
23. Pratt, V.: Action logic and pure induction. In: van Eijck, J. (ed.) JELIA 1990. LNCS, vol. 478, pp. 97–120. Springer, Heidelberg (1991). https://doi.org/10.1007/BFb0018436

Transcendental Syntax IV: Logic Without Systems

Jean-Yves Girard[(✉)]

Directeur de Recherches émérite, Marseille, France
jeanygirard@gmail.com

For André

Abstract. A derealistic, system-free approach, with an example: arithmetic.

Keywords: Logic · Arithmetic · Derealism

1 BHK Revisited

1.1 A System-Free Approach

According to a widespread prejudice, logic should depend upon a system limiting the validity of its laws. Typically, the excluded middle should be accepted in the classical chapel but refused in the intuitionistic bunker. A conception that Kreisel refuted in his day: the polemics as to $A \vee \neg A$ does not concern the system, but the connective, i.e., $\vee := \mathram{?}$ vs. $\vee := \oplus$.

The first evidences against this "fishbowl" view of logic date back to the early 1930's. Typically, Gentzen's *subformula property* which restricts proofs of A to the constituents of A, thus excluding the wider system in which A may have been proved. But the most spectacular blow against bunkerisation is to be found in BHK (Brouwer-Heying-Kolmogoroff), which presents a sort of functional definition of proofs (Sect. 3).

This approach, which does not refer to any system, acknowledges the fact that logic deals with pure reason, truths beyond discussion.

1.2 Axiomatic Realism

Getting rid of systems means standing up against *axiomatic realism*, the duality between syntax and its realistic counterpart, semantics.

But axiomatics and semantics have little to do with proofs. Being concerned with *falsification*, they are, so to speak, scouting the intellectual wilderness: the consistency of $\neg A$ (or the existence of a model refuting A) shows that we shouldn't waste energy in trying to prove A. By telling us *where not to go*, they are very precious auxiliaries, but too warped to be anything more, since they

© Springer Nature Switzerland AG 2020
V. Nigam et al. (Eds.): Scedrov Festschrift, LNCS 12300, pp. 17–36, 2020.
https://doi.org/10.1007/978-3-030-62077-6_2

yield *contingent* truths: A may be valid in system $\mathbb{T}$ and its negation $\neg A$ in system $\mathbb{U}$, both being consistent.

Axiomatics and semantics deal with counterexamples, i.e., impermanence. While our basic interest lies in logic, i.e., permanence.

1.3   The First Leakage: Emptiness

BHK, although the only approach respecting the meaning of the word "logic", has serious leaks. The most obvious being *emptiness*: what to do in presence of formulas with no proofs, typically the absurdity **0** ? Since $\neg A := A \Rightarrow \mathbf{0}$, a proof of a negation becomes a function with the empty set – an unfriendly fellow – as target; this forces the source to be empty as well, in which case the proof becomes the bleak empty function $\emptyset$.

The emptiness of **0** justifies the excluded middle: either A has a proof or it has none, in which case the empty function which maps proofs of A to proofs of **0** is a proof of $\neg A$. This is quite embarrassing and various modifications, none of them definite, have thus been proposed, yielding various *realisability* interpretations. Those "semantics of proofs" are only useful tools, not the real thing – just like a scout is not the Army.

The only way to fix the leakage is to allow all formulas to have proofs, a proposal which conflicts with consistency. Not quite indeed: it is enough to distinguish, among proofs, the real ones from those which are here "to fill the holes". A situation akin to what happens with computer folders: those who look empty to the user indeed harbour "invisible" files .xxx which contain essential informations, the name of the folder or the list of its visible files.

Every proposition, including the absurdity $\mathbf{0} := \mathop{!}(\daleth \,\mathbin{⅋}\, \beth) \otimes \beth$ (Sect. 3.4), admits "proofs". A *truth* criterion (Sect. 3) will determine which ones are visible, i.e., "true"; in the case of absurdity, none.

1.4   The Second Leakage: Operationality

The functions at work in Definition 2 play an essential role, but their status remains rather vague. Should we understand them as computable (recursive) functions or plain set-theoretic graphs? Each answer leads to a specific *category* of morphisms, i.e., a semantics. Categories presuppose the *form* (whence the word "morphism"): their intrinsic essentialism makes them one of the best semantic artifacts, but surely not a way out the bunker.

It seems that rock bottom was hit with the *constellations* of [4], that I will rename *designs*. A product of the experience of proof-theory and computer science, they embody the lessons of Gentzen (their *stars* are sort of logic-free sequents), Herbrand (they socialise through *unification*), logic programming (they look like deterministic PROLOG programs) and proof-nets.

Under certain circumstances, two designs may merge to form a new one through a process that may diverge: this *normalisation* is akin to the traditional cut-elimination – or the *resolution* of logic programming.

This process, which corresponds to the functional application at work in Definition 2, presupposes neither logic nor categories: the merger of two designs can be expressed as a composition... provided we select appropriate sources and targets, but there is no univoque way to do so.

1.5   The Third Leakage: Language

BHK is concerned with those formulas taken from a given language, typically arithmetic. Of course, if we want to free ourselves from systems, we must be ready to consider new formulas and connectives, including eccentric ones, i.e., not limit ourselves to an *a priori* choice: we should be able to consider general propositions, not only those available in a particular fishbowl.

The naive definition of those language-free formulas, called *behaviours*:

A behaviour is any set of designs.

is not technically mature: it must be regulated, typically to exclude the nightmare of emptiness.

The basic example of such a regulation is given by the *correctness criterion* of proof-nets. Which can be expressed in terms of a duality between designs: $\mathcal{P}$, the one under testing vs. $\mathcal{T}$, the test. The test succeeds if the combination $\mathcal{P} + \mathcal{T}$ merges into a design of a certain form, notation

$$\mathcal{P} \perp \mathcal{T}$$

Hence given a set $\mathbf{P}$ of designs, we can define its orthogonal $\sim \mathbf{P}$, i.e., the set of tests it passes. The biorthogonal $\sim\sim \mathbf{P}$ is, so to speak, the regulated version of $\mathbf{P}$, indeed the behaviour generated by $\mathbf{P}$.

Definition 1
A behaviour *is any non trivial set of designs equal to its biorthogonal.*

"Non trivial" means that the behaviour and its orthogonal are non empty. With denumerably many designs, the number of possible behaviours has the power of the continuum. No fishbowl can harbour that many propositions!

1.6   The Fourth Leakage: Usine

This happened to be the only leakage ever observed in the literature. Assuming everything works swell – and it does with our definitions – remains the problem of the distinction between *usine* and *usage* (*factory* and *use*, the use of French emphasising the opposition). L'usage is nothing but the BHK definition, which yields functions, etc. L'usine is the place where we get the certainty that those so-called functions do what they mean to do.

The successful passing of the tests implies cut-elimination and consistency. Therefore incompleteness forbids any form of absolute certainty as to l'usine which usually involves infinitely many tests.

People addressing the issue did not seem to realise that they were up against incompleteness. For instance those asking that, besides the functional proof of Definition 2, one should add an auxiliary proof that the function does what it means to do. But how to deal with this "meta-proof"? If we treat it in the BHK style, it will need in turn its own auxiliary proof, etc.: metas all the way down. In [8], Kreisel proposed to make the meta-proof a formal one in a system given in advance – but later claimed (private communication, circa 1979) that this was a practical joke.

We do know that consistency proofs are impossible, that the Hilbert program cannot be fixed. So let us address the issue without any dogmatism. A behaviour **G** is the orthogonal of a set of tests, a "preorthogonal". The most elementary behaviours admit finite preorthogonals and will therefore be subject to a completely finite checking. But the preorthogonal is, most of the time, infinite and there is no way to implement infinitely many tests: the fact that $\mathcal{F}$ is a BHK proof cannot be an absolute certainty. It can, however, be justified by the usual tools of mathematics, i.e., within set theory.

See annex, p. 18 for further developments.

2 The Architecture of Logic

2.1 Logic vs. Set Theory

We propose to delegate the abstract testing (usine) to set theory: this makes our ultimate – reductionist foundations – depend upon set theory. Just like axiomatic realism, whose justification boils down to some set-theoretic semantics. Both approaches, derealistic and realistic thus rest upon plain mathematics, so let us compare the two approaches in foundational terms.

Set theory is a system, but a well-established one, so flexible and universal that one hardly notices its boundaries: for us, it is mathematics, period. If we insist upon absolute certainty (Sect. 2.3), we must acknowledge the possibility of a failure of this framework. This highly unreasonable occurrence would equally affect both approaches.

Set theory being incomplete, it is likely that it cannot establish that some proof is a proof, i.e., miss the fact that some design $\mathcal{P}$ belongs to some behaviour **G**. But this limitation of the derealistic approach, based on far-reached unprovable statements, is mainly theoretical. On the other hand, the realistic approach is most effectively limited by the walls of its self-chosen prison. As a consequence, the metastatic proliferation of systems.

Take for instance my system $\mathbb{F}$ of fifty years ago [3]: *les candidats de réductibilité* – which are the prefiguration of behaviours – are handled by means of the comprehension principle. If we still see it as a system, we are bound to build extensions – not necessarily bad, like the *constructions* of Coquand [1] –, but sort of prisons anyway. Or we could dump the idea of any system and directly work on behaviours, with almost unlimited possibilities.

Last but not least, most systems are wrong because the semantic justification leaks. The notion is easily tampered with and "bad witnesses" eliminated:

this is what happened to the embarrassing empty model of predicate calculus
(Sect. 2.3).

2.2 Systems vs. Toolbox

So we don't quite need logical systems: if we are not happy with our formulas,
connectives, etc., define new ones by biorthogonality, establish their basic properties and *add them to our data base*. This stock may take the form of an *open*
toolbox containing various designs together with the name of the behaviour they
belong to. A list of untyped artifacts – delogicalised proofs – together with their
types, those types being attributed externally, by arbitrary mathematical methods. The toolbox requires no sophisticated logical structure, e.g., a sequent calculus formulation: we can even use the most archaic logical formulation (axioms
and *Modus Ponens*), which allows us to draw consequences from the principles
listed in the data base, i.e., combine the tools. No cut-elimination, normalisation, etc. at the level of the toolbox is needed, since it is the task of the tools
themselves: when we combine them by *Modus Ponens*, they initiate a converging
merging process.

This is a major improvement over the fishbowl approach for which each
novelty prompted a change of system, the creation of a schismatic chapel. An
approach which culminated with *logical frameworks* [7] where systems $\mathbb{T}, \mathbb{U}, \mathbb{V}, \ldots$
could be put under the same roof with no right to communicate: like hospital
patients, each of them quarantined in his room, lest he contaminate the others.

The fact that l'usine has been delegated to current mathematics, i.e., set
theory, makes our toolbox absolutely faultless – except the legitimate doubt
(Sect. 2.3). The only limit to this approach is our own imagination.

2.3 Certainty

The logical discussions of yesteryear were polluted by the obsession of foundations. We must adopt an adult approach to the question and reflect upon our
certainties or, dually, our doubts.

Generally, the testing cannot actually be performed – it is infinite – and
is delegated to set theory. It is *legitimate* to doubt as to the reliability of set
theory – in the same way we cannot be absolutely confident in the daily return
of the Sun. But these doubts are not quite *reasonable*. Some form of certainty
thus arises from the set theoretic foundation of logic: I call it *epidictic*. Due to
incompleteness, this certainty is only reasonable, not absolute: it leaves some
room for limited, but legitimate, doubts.

The old foundational approach did not distinguish between legitimate and
reasonable: it was seeking a sort of *apodictic* certainty – the one which leaves not
the slightest doubt – and neglected anything irrelevant to this chimeric issue. It
promoted a reductionistic viewpoint based on brute force – consistency as rock
bottom –, thus excluding any sort of finesse.

Like any kind of religious approach, the developments of the apodictic ideology contradict its goals. The search for final justifications leads to overlook

obvious mistakes, for which the doubt is more than legitimate, reasonable: typically the ludicrous principle $\forall \Rightarrow \exists$. Based on the misuse of variables, it is obviously false; but consistent hence, from the apodictic ideology which deals with "strength", a neglectable drawback. The Al Capone method was applied to the embarrassing witness – the empty model – which refutes the nonsense: it was disposed of on the way to Court, this is why models are supposed to be non empty!

2.4   Constraints and Freedom

As we observed with the dubious $\forall \Rightarrow \exists$, each axiomatic system can be justified by means of an *ad hoc* relation to reality. This is precisely why their results are not portable: these systems are prisons, with their own approach to reality, what they call semantics. If we can still use such a prostituted expression, derealism is *the* ultimate semantics.

It is therefore much demanding and does not content itself with a model. For instance, they were serious grounds for the logical constants 1 and $\perp$ of linear logic: no need to explain the interest of having neutral elements for the multiplicative connectives. However, altough a considerable amount of energy was devoted to that peculiar task, the theory of proof-nets never worked for those constants. There is only one way out, namely accept the fact that 1 and $\perp$ are wrong, i.e., impossible. Forcing them to integrate the bulk of logic would destroy the whole architecture. By the way, if we insist upon something of the like, $\forall X \, (X \multimap X)$ and $\exists X \, (X \otimes \sim X)$ will provide reasonable ersatz, but not the real thing which remains a logical fantasy.

The point of good constraints is that they create freedom. Derealism refuses 1 and $\perp$ but accepts equality, the most notorious failure of axiomatic realism, based upon the Leibniz definition

Any property of a *is a property of* b.

As observed in [5], individuals a and b can never be equal, since they can be distinguished by their position w.r.t. "and". Axiomatic realism will object by claiming that we are actually speaking of the respective denotations, i.e., semantics, of these objects and that properties should be consistent with denotations. But how do we know that a property only depends upon the semantics? Elementary, my dear Watson: when it is compatible with... equality! This circular riddle is implemented in various systems telling us which properties are legit. Hence, without system, no Leibniz definition, no equality. By the way, the proof-theoretical treatment of equality is admittedly *ad hoc*: it involves generalised identity axioms embodying the cuts one cannot eliminate, e.g., $t = u, v = u, A[t] \vdash A[v]$.

But who told us that there is a special, segregated category of "individuals" proceeding from the Sky; furthermore that they harbour properties in the same way dogs have a tail? Wouldn't it be simpler if those individuals were just plain propositions, equality being equivalence? This obvious solution can indeed be used to define natural numbers and *prove* the third and fourth Peano axioms (Sect. 5). Exit the aporia of the Leibniz equality.

So why did it take so long to integrate the most natural logical primitive? Simply because of the classical prejudice: up to consistency, everything is classical, hence the excluded middle

$$A \equiv B \ \vee \ B \equiv C \ \vee \ C \equiv A \tag{1}$$

which implies the impossibility of three unequivalent propositions. Intuitionism, which does not agree on this, does not disagree either, i.e., proves $\neg\neg$ (1). Linear logic – which should not be seen as a system, but a space of freedom –, by restricting the contraction rule to specific cases, makes (1) the exception, by no means the rule. No doubt a useful exception, but which can be a pain in the neck in some cases.

Another issue related to freedom: the paper [6] introduced *light exponentials*, i.e., connectives dedicated to perenniality, with some relation to computational complexity. They were developed in various systems (BLL, LLL, ELL...) whose relative qualities I shall not discuss for the very reason that we move on sort of quicksand, with no real benchmark: the semantics turns out to be more treacherous than ever. This is why it would be of utmost importance to determine whether or not light exponentials are more than a figment of axiomatic realism, in other terms whether they can be explained in terms of behaviours.

3 Truth

3.1 The Tarskian Pleonasm

It suffices to compare BHK

Definition 2

A proof of $A \Rightarrow B$ is a function $\mathcal{F}$ mapping any proof $\mathcal{P}$ of A to a proof $\mathcal{F}(\mathcal{P})$
of B.

to Tarski's "definition" of truth, e.g.,

Definition 3
$A \Rightarrow B$ *is true when the truth of A implies the truth of B.*

(and its declinations for $\wedge, \vee, \neg$,... in terms of and, or, not, ...) to see the difference between an inspired approach and a pleonasm which boils down to "*A* is true when *A*". But the truism is the ultimate form of snobbery: you think the Emperor is naked, mistake, you just don't see his new clothes.

Indeed, the famous *vérité de La Palice*, a theory of truth due to a French precursor of analytic philosophy, e.g.,

> *Un quart d'heure avant sa mort, il était encore en vie.*

foreshadows Definition 3.

The current opinion among non believers is that tarskian truth is, unfortunately, correct. But even this correctness is dubious, since truth does not apply to formulas but to proofs! Sect. 4.3 wil provide us with examples *contradicting* the tarskian definition, which is thus not even a pleonasm.

3.2 Generalities About Visibility

Remember that we definitely dumped fishbowls, hence no longer deal with the formulas of a language, but with general behaviours (Definition 1). Our definition of truth takes the form:

Definition 4
G *is true when it harbours a visible design.*

The visible designs are the true ones, the actual *proofs* so to speak. Visibility, yet to be defined, should enjoy certain implicit requirements:

- It should be closed under cut: hence, if $\mathcal{P}$ and $\mathcal{F}$ are proofs of **G** and **G** $\Rightarrow$ **H**, then the design $\mathcal{F}(\mathcal{P})$ of **H** must be visible, i.e., a proof of **H**.
- Some behaviour, typically the absurdity **0**, must be without visible element, i.e., not true.

If these requirements are satisfied, then truth is consistent: **G** and its classical negation **G** $\Rightarrow$ **0** cannot both have visible designs, i.e., both be true. An exclusion that does not extend to linear negation: the self-dual behaviours $\daleth = \sim\daleth$ and $\beth = \sim\beth$ are true.

Since truth deals with proofs and not with mere provability, the truth of a compound behaviour cannot be reduced to the truth of its constituents. Therefore it cannot follow any kind of truth table. In particular, a conjunction may be true while one of the conjuncts is not. So tarskian truth is worse than a useless and snobbish ready-made, it is a plain mistake!

3.3 Multiplicative Case

We shall first explain the solution in the case of the multiplicative proof-nets of linear logic; we consider formulas built from literals $p, \sim p, q, \sim q, r, \sim r, \ldots$ by means of $\otimes$ and $\invamp$. Besides the usual $\otimes$ and $\invamp$-links, we allow arbitrary links $\overbrace{p_1, \ldots, p_k}$, with $k > 0$, which resemble axioms in the sense that they are without premise. The usual correctness criterion is applied to the structures built from those, $\overbrace{p_1, \ldots, p_k}$ being seen as a vertex with edges $p_1, \ldots, p_k$: this generalises the usual case based on the sole $\overbrace{p, \sim p}$, see Sect. 3.5 below.

A proof structure with literals $q_1, \ldots, q_N$ (with possible repetitions, this is the familiar nonsense about "occurrences") can be seen as a partition $\mathcal{P}$ of $\{1, \ldots, N\}$ the classes of which are precisely the "axioms" $\overbrace{q_{i_1}, \ldots, q_{i_k}}$ used. A switching of the proof-net yields another partition $\mathcal{T}$ of the same $\{1, \ldots, N\}$. Both partitions can be put together to form a bipartite graphs: the classes being its vertices, an edge between $X \in \mathcal{P}$ and $Y \in \mathcal{T}$ is an element of $\mathcal{P} \cap \mathcal{T}$. The Danos-Regnier criterion [2] requires that, for any $\mathcal{T}$ arising from a switching, the induced graph is connected and acyclic. In particular $X \in \mathcal{P}$ and $Y \in \mathcal{T}$ intersect in at most one point.

Let n and m be the respective numbers of partitions in $\mathcal{P}$ and $\mathcal{T}$. If the proof is correct, then $n + m - N = 1$ (Euler-Poincaré), what can be written

$(2n - N) + (2m - N) = 2$. The *weight* $|\mathcal{P}| := 2n - N$, which does not depend upon $\mathcal{T}$, can be written as the sum of the weights of its "axiom links" defined by $|\overbrace{p_1, \ldots, p_k}| = 2 - k$. Our visibility definition writes as:

Definition 5
$\mathcal{P}$ *is visible when* $|\mathcal{P}| \geq 0$.

Observe that the weight of the familiar $\overbrace{p, \sim p}$ is $2 - 2 = 0$, hence a proof-net using the familiar identity axioms is of total weight 0, hence visible.

Visibility satisfies the requirements of the previous section. First, it is deductively closed: normalising a cut amounts at replacing $\overbrace{p_1, \ldots, p_k, p}$ and $\overbrace{\sim p, q_1, \ldots, q_\ell}$ (total weight $2 - (k + 1) + 2 - (\ell + 1) = 2 - (k + \ell)$) with $\overbrace{p_1, \ldots, p_k, q_1, \ldots, q_\ell}$ (weight $2 - (k + \ell)$, i.e., the same). Moreover, not everything is true: typically, $p \,⅋\, q \,⅋\, r$, whose only correct proof-net, which uses the axiom $\overbrace{p, q, r}$ of weight -1, is invisible.

Incidentally, we gave the fatal blow to tarskian truth: $(p \,⅋\, q \,⅋\, r) \otimes s$ is true while $p \,⅋\, q \,⅋\, r$ is not.

3.4   The Constants Are Dead, Long Live the Constants!

Our multiplicative example has been oversimplified for pedagogic purposes. Atoms indeed split into two classes, *objective* and *subjective*, each one being closed under negation. This modification makes it possible to handle the absurdity **0** and is the key to second order (Sect. 5.2). It only affects the weighing of "axioms", written $\overbrace{p_1, \ldots, p_k, q_1, \ldots, q_\ell}$, the p_i being objective, the q_j subjective.

– If $\ell = 0$, i.e., if the axiom is objective, then $|\overbrace{p_1, \ldots, p_k}| = 2 - k$.
– Otherwise, $|\overbrace{p_1, \ldots, p_k, q_1, \ldots, q_\ell}| = -k$.

Subjective atoms, whatever their number, count for two objective ones.

Keeping Definition 5 of visibility, it remains to show the deductive closure of truth. $|\overbrace{p_1, \ldots, p_k, q_1, \ldots, q_\ell, p}|$ takes the value $-k$ if p is subjective; if p is objective, it takes one of the values $2 - (k + 1)$ (if $\ell = 0$) and $-(k + 1)$ (if $\ell \neq 0$). Ditto with $|\overbrace{\sim p, r_1, \ldots, r_{k'}, s_1, \ldots, s_{\ell'}}|$: possible weights $-k'$, $2 - (k' + 1)$ and $-(k' + 1)$. Both of them weight $a - (k + k')$ where a takes one of the values $2, 0, -1, -2$: $a = 1$ is excluded since this would require, say, p to be objective and $\sim p$ subjective. On the other hand, $\overbrace{p_1, \ldots, p_k, q_1, \ldots, q_\ell, r_1, \ldots, r_{k'}, s_1, \ldots, s_{\ell'}}$ weights $2 - (k + k')$ or $-(k + k')$. The weight can decrease during normalisation only if $a = 2$, in case $\ell = \ell' = 0$ but the normalised "axiom" would weight $2 - (k + k')$.

Indeed, up to linear equivalence, there are only two atoms, the objective ⼀ ("fu") and the subjective ⼄ ("wo"). Both are true, self-dual and inequivalent. They can be used to define the absurdity by $\mathbf{0} := !(⼀ ⅋ ⼄) \otimes ⼄$. Indeed, Sect. 4.2 of [5], proves, without using the notations (⼀ and ⼄ were still in limbo) the rule

$$\frac{\vdash \Gamma, A}{\vdash \Gamma, \top}$$

which is an alternative formulation of the famous *ex nihilo quod libet* $\mathbf{0} \multimap A$. Incidentally, the notion of *épure* (= working drawing) of paper [5] is different: either $k = 2, \ell = 0$ or $k = 0$. This ensures that $|\overbrace{p_1, \ldots, p_k, q_1, \ldots, q_\ell}| = 0$, hence épures are visible.

The constants ($\mathbf{1}$ and $\bot$) are dead, long live the constants ($\daleth$ and $\beth$). Whose multiplicative combinations yield natural numbers (Sect. 4 below).

3.5 Variables

According to a dubious tradition, propositional calculus should be built from unspecified constants $P, Q, R, \ldots$. Weird constants indeed, for which anything can be substituted: this is what one usually calls variables! But such variables should then be styled *second order*, a part of logic against which a fatwa was declared. Let us call a spade a spade and use the notation $X, Y, Z, \ldots$ to emphasise the fact that we are dealing with variables.

Those variables were part of proof-nets original, since we needed some sorts of atoms. Those proof-nets made use of binary identity links $\overbrace{X, \sim X}$. They are compatible with our truth definition, since they are binary; their weight is always zero, since X and $\sim X$ are simultaneously objective or subjective.

The restriction to the links $\overbrace{X, \sim X}$ has nothing to do with a sort of systemic ukase, it can be derived from closure under instanciation: the net should remain correct when we replace its variables with independent propositions. This can take the form of a switching (already presented in [4], but without the notation $\daleth$), involving the selection of a "value" for each variable X with three cases:

$$X := \daleth \qquad \sim X := \daleth$$
$$X := \daleth \otimes \daleth \sim X := \daleth \,\invamp\, \daleth$$
$$X := \daleth \,\invamp\, \daleth \sim X := \daleth \otimes \daleth$$

This excludes all possible practical jokes, e.g., $\overbrace{X}$, $\overbrace{X, Y}$, $\overbrace{X, X}$, $\overbrace{X, X, \sim X}$.

3.6 General Case

We are not quite dealing with proof-nets, but with the designs of a behaviour. The main difference with the multiplicative case is that duplications and erasings may occur during normalisation. Our numerical criterion is obviously sensitive to these operations, hence we must be cautious.

The truth of $\mathcal{P} \in \mathbf{G}$ is related to the testing process. So let $\mathcal{T}$ be a test in $\sim \mathbf{G}$, hence $\mathcal{P} \perp \mathcal{T}$. The actual performance of the test, a normalisation in the sense of [4], involves the building of a connected-acyclic graph whose vertices are made of two designs, $\mathcal{P}_\mathcal{T}$ and $\mathcal{T}'$, each ray of $\mathcal{P}_\mathcal{T}$ being a ray of $\mathcal{T}'$; the edges are those common rays. $\mathcal{P}_\mathcal{T}$ and $\mathcal{T}'$ are obtained through a unification (matching) procedure which replaces any star σ of those designs with various substitutions $\sigma\theta_i$.

The visibility of $\mathcal{P}$ is obtained by means of a weighing of the stars of $\mathcal{P}_\mathcal{T}$. Remembering that rays are divided into objective and subjective ones, let $[\![t_1, \ldots, t_k, u_1, \ldots, u_\ell]\!] \in \mathcal{P}_\mathcal{T}$, then:

- If $\ell = 0$, then $|[\![t_1, \ldots, t_k]\!]| = 2 - k$.
- Otherwise, $|[\![t_1, \ldots, t_k, u_1, \ldots, u_\ell]\!]| = -k$.

The closure of visibility under cut is the consequence of the fact that the matching between t and u of complementary colours is impossible if one is objective and the other subjective. Generally, the testing should anticipate general normalisation; in terms of truth, it should make sure that the $\mathcal{P}_\mathcal{T}$ are representative of the $\mathcal{P}_\mathcal{Q}$ occurring during the actual normalisation of a cut between $\mathcal{P}$ and a design $\mathcal{Q}$, visible or not, in some $\vdash \sim \mathbf{G}, \mathbf{\Gamma}$.

4   Natural Numbers

We now restrict our attention to the multiplicative combinations of the self-dual constants $\daleth$ and $\gimel$. We shall classify them up to linear equivalence (i.e., logical equality) $A \equiv B := (A \multimap B) \& (B \multimap A)$.

4.1   First Series

Definition 6
The weight of the multiplicative A built from the sole $\daleth$ is defined as the common weight of all designs of A:

$$|\daleth| = 1$$
$$|A \otimes B| = |A| + |B|$$
$$|A \,\gimel\, B| = |A| + |B| - 2$$

In particular, $|\sim A| = 2 - |A|$ and $|A \multimap B| = |B| - |A|$.

For $n > 0$ define $\daleth_n := \daleth \otimes \daleth \otimes \ldots \otimes \daleth$ (a n-ary tensor) and for $n < 2$ $\daleth_n := \daleth \,\gimel\, \daleth \,\gimel\, \ldots \,\gimel\, \daleth$ (a $2 - n$-ary par), both cases agreeing on $\daleth_1 := \daleth$. Observe that $\sim \daleth_n \equiv \daleth_{2-n}$.

Theorem 1
$A \equiv \daleth_{|A|}$

Proof: By recurrence on the size of A, the basic case $A = \daleth$ being trivial. If A is a tensor $B \otimes C$ and $B \equiv \daleth_{|B|}, C \equiv \daleth_{|C|}$, then $A \equiv \daleth_{|B|} \otimes \daleth_{|C|} \equiv \daleth_{|A|}$. If A is a "par" $B \,\gimel\, C$, the previous case shows that $\sim A$ is equivalent to $\daleth_{|\sim A|}$, hence $A \equiv \sim \daleth_{2-|A|} \equiv \daleth_{|A|}$.   □

$\daleth_0 = \daleth \,⅋\, \daleth$ is a sort of corrected version of the late neutral $\mathbf{1}$, ditto for $\daleth_2 = \daleth \otimes \daleth$ w.r.t. $\bot$. $\daleth_0$ and $\daleth_2$ are, so to speak, the logical part of the multiplicative units. They basically work because $\daleth$ and $\exists$ no longer follow any semantic paradigm!

All $\daleth_n$, for $n \geq 0$, are provable. As a particular case, $\daleth_0$, $\daleth_1$ and $\daleth_2$ are provable together with their linear negations $\daleth_2$, $\daleth_1$ and $\daleth_0$. For $n < 0$, the $\daleth_n$ are not provable; they are indeed *refutable* (Sect. 4.3).

4.2   Second Series

For $n \in \mathbb{Z}$, we define the $\exists_n$: $\exists_0 := \exists$, $\exists_n := \daleth_n \otimes \exists$ when $n \neq 0$.

Proposition 1

$$\exists \equiv \daleth_0 \otimes \exists$$

Proof: From $\vdash \exists, \exists$ and $\vdash \daleth_0$, we get $\vdash \exists, \daleth_0 \otimes \exists$, hence the implication $\exists \multimap \daleth_0 \otimes \exists$. Conversely, $\vdash \daleth$ and $\vdash \daleth, \exists, \exists$ admit designs of respective weights 1 and -1 which combine into a design of weight 0 of $\vdash \daleth \otimes \daleth, \exists, \exists$ which yields a proof of $\daleth_0 \otimes \exists \multimap \exists$.   □

Hence $\exists_n \equiv \daleth_n \otimes \exists$ for all $n \in \mathbb{Z}$. More generally:

Proposition 2

$$\daleth_m \otimes \daleth_n \equiv \exists_{m+n}$$

(obvious) and

Proposition 3

$$\exists_m \otimes \exists_n \equiv \exists_{m+n}$$

Proof: Boils down to $\exists \otimes \exists \equiv \exists$. From $\vdash \exists, \exists, \exists$, we get $\exists \otimes \exists \multimap \exists$; conversely, $\vdash \exists, \exists$ and $\vdash \exists$ yield $\vdash \exists, \exists \otimes \exists$, hence $\exists \multimap \exists \otimes \exists$.   □

Proposition 4

$$\daleth_{n+2} \,⅋\, \exists \equiv \exists_n$$

Proof: From $\vdash \exists, \exists$ and designs in $\sim\daleth_{n+2}$ and $\daleth_n$ of respective weights $-n$ and n, one gets a proof of $\vdash (\sim\daleth_{n+2} \otimes \exists), (\daleth_n \otimes \exists)$, hence $\daleth_{n+2} \,⅋\, \exists \multimap \exists_n$. Conversely, from $\vdash \exists, \daleth_2, \exists$, we get $\vdash \sim\daleth_n \,⅋\, \exists, (\daleth_2 \otimes \daleth_n) \,⅋\, \exists$, hence $\exists_n \multimap \daleth_{n+2} \,⅋\, \exists$.   □

Theorem 2

Any multiplicative combination A of $\daleth$ and at least one $\exists$ is provably equivalent to some $\exists_n$.

Proof: By recurrence on the size of A, the basic case $A = \exists$ being trivial. If A is a tensor $B \otimes C$, at least one of B and C uses a $\exists$ and we are left with the cases $\exists_m \otimes \daleth_n$, $\daleth_m \otimes \exists_n$ and $\exists_m \otimes \exists_n$ which by Propositions 2 and 3 are equivalent to $\exists_{m+n}$. If A is a "par" $B \,⅋\, C$, the previous case shows that $\sim A$ is equivalent to some $\exists_n$, hence $A \equiv \daleth_n \,⅋\, \exists$; using Proposition 4, we get $A \equiv \exists_{n-2}$.   □

Definition 6 can be extended to multiplicative combinations of $\daleth$ and $\Game$:

Definition 7

$$|\daleth| = 1$$
$$|\Game| = 0$$
$$|A \otimes B| = |A| + |B|$$
$$|A \,⅋\, B| = |A| + |B| - 2 \quad \text{if one of } A, B \text{ is } \Game\text{-free}$$
$$|A \,⅋\, B| = |A| + |B| \quad \text{otherwise}$$

By 1 and 2, A is equivalent to either $\daleth_{|A|}$ or $\Game_{|A|}$. In general:

1. $\daleth_m \otimes \daleth_n \equiv \daleth_{m+n}$ and $\daleth_m \,⅋\, \daleth_n \equiv \daleth_{m+n-2}$
2. $\sim\daleth_n \equiv \daleth_{2-n}$ and $\daleth_m \multimap \daleth_n \equiv \daleth_{n-m}$
3. $\Game_m \otimes \Game_n \equiv \Game_m \,⅋\, \Game_n \equiv \Game_{m+n}$
4. $\sim\Game_n \equiv \Game_{-n}$ and $\Game_m \multimap \Game_n \equiv \Game_{n-m}$
5. $\Game_m \otimes \daleth_n \equiv \Game_{m+n}$ and $\Game_m \,⅋\, \daleth_n \equiv \Game_{m+n-2}$
6. $\Game_m \multimap \daleth_n \equiv \Game_{n-m-2}$ and $\daleth_m \multimap \Game_n \equiv \Game_{n-m}$

4.3   Truth and Falsity

Theorem 3
The $\daleth_n$ and $\Game_n$ are refutable for $n < 0$.

Proof: $\daleth_n \multimap \Game_n$ being equivalent to $\Game_0 \,(= \Game)$, is provable; $\neg\Game_n \multimap \neg\daleth_n$ is thus provable, which reduces the theorem to the case of $\Game_n$. Now $\Game_n \multimap \Game_{-1}$ being equivalent to the provable $\Game_{-1-n}$, we are reduced to proving $\neg\Game_{-1}$: from $\vdash \sim\Game_{-1}$, $\Game_{-1}$ and $\vdash \daleth$, we get $\Game_{-1} \Rightarrow\, !\Game_{-1} \otimes \daleth$, i.e., $\Game_{-1} \Rightarrow \mathbf{0}$ that is the negation $\neg\Game_{-1}$.   □

Let us sum up the basic facts about truth and falsity (i.e., refutability):

1. The $\daleth_n$ and $\Game_n$ are true for $n \geq 0$, false for $n < 0$.
2. The implications $\daleth_m \multimap \daleth_n$, $\daleth_m \multimap \Game_n$ and $\Game_m \multimap \Game_n$ are true for $m \leq n$, false when $m > n$.
3. The implication $\Game_m \multimap \daleth_n$ is true when $n \geq m+2$, false otherwise.

The two series are thus distinct, the sole relation between them being the double implication

$$\daleth_n \multimap \Game_n \multimap \daleth_{n+2}$$

We definitely *contradict* the excluded middle (1) which forbids the existence of three provably unequivalent propositions! This implies necessary divergences from classical truth which are made possible by the fact that our truth applies to proofs and not to propositions. In particular the novelty cannot be tamed by a change of truth tables, say replacing $\mathbf{t},\mathbf{f}$ with $\mathbb{Z}$. Typically, A of weight n can be equivalent to $\daleth_n$ or $\Game_n$.

The following table is a list of possible deviations (with $\mathbf{t}$ = true, $\mathbf{f}$= false) w.r.t. classical truth. The first line, with $A = B = \daleth_0$, yields $A \,⅋\, B = \daleth_{-2}$ and $\sim A = \daleth_2$. The second line, with $A = \daleth_{-1}, B = \daleth_1$, yields $A \otimes B = \daleth_0$ and $A \,⅋\, B = \daleth_{-2}$. No deviation when both A and B are false. "$⅋$" is more deviant than "$\otimes$": this is because negation does not exchange $\mathbf{t}$ and $\mathbf{f}$.

A B	$A \otimes B$	$A \,⅋\, B$	$\sim A$
$\mathbf{t}$ $\mathbf{t}$		$\mathbf{f}$	$\mathbf{t}$
$\mathbf{f}$ $\mathbf{t}$	$\mathbf{t}$	$\mathbf{f}$	

A definite jailbreak from tarskism... and any sort of semantics.

5   Arithmetic

We shall now reconstruct arithmetic, not as a system, but as part of our open logic. We basically need two sorts of quantifiers, first and second order.

5.1   First Order Quantification

First order quantification is about relative numbers, identified with the series $\daleth_n$ ($n \in \mathbb{Z}$). The following can serve as a definition of *individuals*:

1. The variables $\mathbf{x}, \mathbf{y}, \mathbf{z}, \dots$ are individuals.
2. $\overline{\mathbf{1}} := \daleth$ is an individual.
3. If $\mathbf{t}, \mathbf{u}$ are individuals, so are $\mathbf{t} + \mathbf{u} := t \otimes u$ and $\mathbf{t} - \mathbf{u} := t \multimap u$.

Since logic is open, we don't even require that (1)–(3) be the only way to build individuals.

The usual rules of quantification do apply, provided we *declare* our variables. Incidentally, due to the presence of the closed individual $\daleth$, the principle $\forall \multimap \exists$ holds: $\forall \mathbf{x} A[x] \multimap A[\daleth]$ and $A[\daleth] \multimap \exists \mathbf{x} A[x]$.

Variables indeed stand for arbitrary behaviours, analogous to the so-called propositional "constants", indeed variables, of logic. The basic parametric proposition (i.e., predicate) is inequality:

$$\mathbf{t} \le \mathbf{u} \quad := \quad \mathbf{t} \multimap \mathbf{u}$$

From which we can define equality:
$$\mathbf{t} = \mathbf{u} \quad := \quad (\mathbf{t} \multimap \mathbf{u}) \,\&\, (\mathbf{u} \multimap \mathbf{t})$$

The standard principles of linear logic allow us to establish certain principles which are usally handled via axiomatics. Typically:

$$\mathbf{x} \le \mathbf{x} \tag{2}$$
$$\mathbf{x} + (\mathbf{y} + \mathbf{z}) = (\mathbf{x} + \mathbf{y}) + \mathbf{z} \tag{3}$$
$$\mathbf{x} + (\mathbf{y} - \mathbf{z}) \le (\mathbf{x} + \mathbf{y}) - \mathbf{z} \tag{4}$$

Let $\overline{0} := 7 - 7 (= 7 \,⅋\, 7)$. Since individuals deal with relative numbers, the third Peano axiom takes the form:

$$\overline{0} \leq x \quad \multimap \quad (x + 7) \neq \overline{0} \tag{5}$$

Which can be proved as follows: from $\overline{0} \leq x$ we get $\overline{0} + 7 \leq x + 7$, which, combined with $(x + 7) = \overline{0}$, yields the refutable $\overline{0} + 7 \leq \overline{0}$.

As to the fourth Peano axiom, the best we can get is the following:

$$(x + 7) \leq (y + 7) \quad \multimap \quad x \leq ((y + 7) - 7) \tag{6}$$

which makes use of $x \multimap ((x + 7) - 7)$. The implication $((x + 7) - 7) \multimap x$ is missing; it is however provable when x is a "successor":

Theorem 4

$$((x + 7 + 7) - 7) \leq x + 7$$

Proof: $\vdash \sim x, x$ and $\vdash 7, 7, 7$ (weight -1) yield $\vdash \sim x \,⅋\, 7 \,⅋\, 7, x \otimes 7$ (weight -1) hence, with $\vdash 7$ (weight 1), $\vdash (\sim x \,⅋\, 7 \,⅋\, 7) \otimes 7, x \otimes 7$.   □

Summing up, we conclude that the fourth Peano axiom holds for successors, so to speak $SSx = SSy \multimap Sx = Sy$.

In terms of proof-nets, universal quantification is handled, as in [5], by means of a switching choosing independent values for the variables: the choices $x = 7, x = 7 \,⅋\, 7, x = 7 \otimes 7$ are enough (Sect. 3.5 *supra*).

Existential quantification is handled as in [5], with a major simplification: the existential witnesses $\mathcal{G} + \tilde{\mathcal{G}}$ were defined as linear combinations of all elements in the *finite* $\mathcal{G}$ and $\tilde{\mathcal{G}}$. We simplify our construction by using, instead of the full t and $\sim t$ a specific test in each of them. With two consequences:

– We no longer use linear combinations (good riddance!).
– The same simplification can be used in the second order case where behaviours are infinite.

First-order is basically weaker than the usual first order of Peano, who could use axiomatics to decide which primitive is legal or not or which principle is true. Since we are concerned with logic, we have no longer access to ukases and are unable to establish the full fourth Peano axiom or define the product $t \cdot u$. The missing "axiom" is trivially proved under the form $x \in \mathbb{N} \multimap ((x + 7) - 7) \multimap x$ by a recurrence (Sect. 5.3), a second order principle[1], just as the missing product is second order definable (Sect. 5.4).

By the way, one of the blind spots of BHK was the handling of purely universal statements of arithmetic. Basically a proof of $\forall x A[x]$ is treated pointwise as a function mapping $n \in \mathbb{N}$ to a proof of $A[\overline{n}]$, which, being a plain computation, can be described in advance, hence the "proof" reduces to the "meta-proof" of Sect. 1.6, which in turn reduces to meta-meta-proofs all the way down. Observe that $((\overline{n} + 7) - 7) \multimap \overline{n}$ holds pointwise but that the proofs do not proceed from a common design; there is indeed one for $n > 0$ which does not merge with the case $n = 0$.

---

[1] Which also proves $x \in \mathbb{N} \multimap \overline{0} \leq x$, hence $x \in \mathbb{N} \multimap (x + 7) \neq \overline{0}$.

5.2 Second Order Propositional Case

Although there is no use for it, let us start with second order propositional quantification, i.e., system $\mathbb{F}$. This was the stumbling block of [5], due to the fact that behaviours are usually infinite: we cannot encapsulate an infinite set inside a design. By the way, should we attempt such a nonsense, we would enter into a wild goose chase as to the cardinality of behaviours.

The original treatment of $\mathbb{F}$ [3] involved *candidats de réductibilité*, which suggests the following definition.

Definition 8

A candidate of base $\mathcal{T} + \mathcal{U}$, where $\mathcal{T}, \mathcal{U}$ are orthogonal tests, is any behaviour **G** such that $\mathcal{T} \in \mathbf{G}$ and $\mathcal{U} \in \sim\mathbf{G}$.

Existential quantification is handled as follows:

Analytically: The *proof* of $\exists\mathbf{X}\mathbf{A}[\mathbf{X}]$ obtained from $\mathbf{A}[\mathbf{T}]$ makes use of a *witness* $\mathcal{T} + \mathcal{U}$, namely the base of the behaviour **T**, seen as a candidate.
Synthetically: The *behaviour* $\exists\mathbf{X}\mathbf{A}[\mathbf{X}]$ is defined by:

$$\exists\mathbf{X}\mathbf{A}[\mathbf{X}] := \sim\sim(\bigcup_{\mathbf{T}}\mathbf{A}[\mathbf{T}]) \tag{7}$$

Our choice of witness is basically a simplification of what we proposed in [5]: since there is no hope of packing together the full $\mathbf{T}, \sim\mathbf{T}$, we cannot avoid partiality (Sect. 6.1 of [5]). Singling out elements $\mathcal{T} \in \mathbf{T}$ and $\mathcal{U} \in \sim\mathbf{T}$ makes it even more partial, but this partiality matters no more in the context of infinite behaviours. Incidentally observe that the existential case actually defines a behaviour: the practical joke of an empty orthogonal is avoided, since it contains the switching $[\![\overline{p_\alpha(\mathrm{mag}(x d y))|p_{\tilde{\alpha}}(\mathrm{mag}(x d y))}]\!]$ ([5], Sect. 6.1) which checks the orthogonality of the pillars $\mathcal{T}, \mathcal{U}$ of the base.

Universal quantification is handled by a plain intersection:

$$\forall\mathbf{X}\mathbf{A}[\mathbf{X}] := \bigcap_{\mathbf{T}}\mathbf{A}[\mathbf{T}] \tag{8}$$

Definitions (7) and (8) follow the original pattern used for system $\mathbb{F}$ ([3]) which now yields a justification of second order principles.

5.3 Recurrence

The principle of recurrence, a.k.a. mathematical induction is usually written:

$$\forall\mathbf{y}\,(A[\mathbf{y}] \multimap A[\mathbf{y} + \daleth]) \Rightarrow (A[\overline{0}] \multimap A[\mathbf{x}]) \tag{9}$$

with two defects, one being that it is an axiom, i.e., an ukase proceeding from the Sky, the other being that it is a schema, i.e., a sort of meta-axiom introduced in order to circumvent the fatwa against second order. Replacing the schema with the obvious second order definition makes it possible to *define* natural numbers, Dedekind style, as the smallest set containing zero and closed under successor:

$$\mathbf{x} \in \mathbb{N} \quad := \quad \forall X \, (\forall \mathbf{y} \, (X(\mathbf{y}) \multimap X(\mathbf{y} + \daleth)) \Rightarrow (X(\overline{0}) \multimap X(\mathbf{x}))) \qquad (10)$$

From which the implication $x \in \mathbb{N} \multimap$ (9) follows. A useful variant is obtained by applying (9) to $!A[x] \otimes (A[x] \multimap A[x])$, which yields:

$$\mathbf{x} \in \mathbb{N} \multimap \forall \mathbf{y} \, (A[\mathbf{y}] \Rightarrow A[\mathbf{y} + \daleth]) \Rightarrow (A[\overline{0}] \Rightarrow A[\mathbf{x}]) \qquad (11)$$

The handling of quantification over predicates, here unary, is inspired from the propositional case. We should introduce a notion of parametric candidate. First by separating positive from negative occurrences. Typically, $\mathbf{x} \multimap \mathbf{x}$ should be written as $\mathbf{x}^- \multimap \mathbf{x}^+$ and later subject to the constraint $\mathbf{x}^- = \mathbf{x}^+$. In terms of parametric candidates, this means that we should consider doubly indexed families $\mathcal{G}_{m,n}$ $(m, n \in \mathbb{Z})$ of candidates enjoying:

$$m' \leq m, n \leq n' \Rightarrow \mathcal{G}_{m,n} \subset \mathcal{G}_{m',n'} \qquad (12)$$

i.e., covariant in n, contravariant in m; the negation will thus be covariant in m, contravariant in n. They should also be provided with a base $\mathcal{T} + \mathcal{U}$ such that, for all $m, n \in \mathbb{Z}$, $\mathcal{T}(\sim \overline{m}, \overline{n}) + \mathcal{U}(\overline{m}, \sim \overline{n})$ is a base of $\mathcal{G}_{m,n}$. Typically, if $\mathcal{G}_{m,n} := \overline{m} \multimap \overline{n}$, $\mathcal{T}$ and $\mathcal{U}$ stand for switchings of $\gamma\!\!\!\gamma$ and its dual $\otimes$, so that $\mathcal{T}(\sim \overline{m}, \overline{n})$ and $\mathcal{U}(\overline{m}, \sim \overline{n})$ are switchings of $\overline{m} \multimap \overline{n}$ and $\overline{m} \otimes \sim \overline{n}$.

5.4 Product

The product $(\mathbf{t} \cdot \mathbf{x}) \simeq \mathbf{y}$ is defined by a quantification over binary predicates:

$$\forall X \, (\forall \mathbf{x} \forall \mathbf{y} \, (X(\mathbf{x}, \mathbf{y}) \multimap X(\mathbf{x} + \daleth, \mathbf{y} + \mathbf{t})) \multimap (X(\overline{0}, \overline{0}) \multimap X(\mathbf{x}, \mathbf{y}))) \qquad (13)$$

We can then prove the existence of the product by recurrence on $\mathbf{x}$:

$$\mathbf{x} \in \mathbb{N} \Rightarrow \exists \mathbf{y} \, (!(\mathbf{y} \in \mathbb{N}) \otimes (\mathbf{t} \cdot \mathbf{x}) \simeq \mathbf{y}) \qquad (14)$$

The predicate $(\mathbf{t} \cdot \mathbf{x}) \simeq \mathbf{y}$ is handled by means of a sort of graph recurrence, which amounts at replacing the variable X of definition (13) with a specific binary predicate $A[\mathbf{x}, \mathbf{y}]$. For instance, with $A[\mathbf{x}, \mathbf{y}] := \mathbf{x} \in \mathbb{N}$, we get:

$$(\mathbf{t} \cdot \mathbf{x}) \simeq \mathbf{y} \multimap \mathbf{x} \in \mathbb{N} \qquad (15)$$

Consider $A[\mathbf{x}, \mathbf{y}, \mathbf{x}', \mathbf{y}'] := \mathbf{x} = \mathbf{x}' \multimap \mathbf{y} = \mathbf{y}'$; the following are provable:

$$A[\overline{0}, \overline{0}, \overline{0}, \overline{0}] \qquad (16)$$

$$A[\overline{0}, \overline{0}, \mathbf{x}' + \daleth, \mathbf{y}' + \mathbf{t}] \qquad (17)$$

$$A[\mathbf{x} + \daleth, \mathbf{y} + \mathbf{t}, \overline{0}, \overline{0}] \qquad (18)$$

$$A[x, y, x', y'] \multimap A[x + \daleth, y + t, x' + \daleth, y' + t] \tag{19}$$

A "graph recurrence" w.r.t. x', y', using (16) and (17) yields

$$(t \cdot x') \simeq y' \multimap A[\overline{0}, \overline{0}, x', y'] \tag{20}$$

Another "graph recurrence" w.r.t. x', y', using (18) and (19) yields

$$(t \cdot x') \simeq y' \multimap (A[x, y, x', y'] \multimap A[x + t, y + t, x', y']) \tag{21}$$

And a graph recurrence w.r.t. x, y, using (20) and (21) yields:

$$(t \cdot x) \simeq y \multimap ((t \cdot x') \simeq y' \multimap A[x, y, x', y']) \tag{22}$$

in other terms, the unicity of the product.

Incidentally, the fact that the product is only second order definable may be related to the typical second order feature known as the incompleteness of arithmetic, which relies on an encoding making a heavy use of the product.

A   L'usine, Again

Usine vs. usage, it's Church vs Curry. The existentialist approach of Curry is quite respected by the notion of *behaviour*. On the other hand, the essentialism inherent to the typing à la Church leads to systems and must be deeply modified. I propose the following:

A type (Church style) is a (finite) battery of tests.

This is compatible with polymorphism: several batteries may be used to "type" the same design. However, there is a problem, the definition seeming not to apply in full generality, because of the absence of finite preorthogonals.

I propose the following solution: instead of a preorthognal of behaviour **P**, a preorthogonal of a *sub-behaviour* of **P**. Orthogonality to such a preorthogonal need not be necessary, it is only *sufficient*. On the other hand, it may remain finite, hence the possibility of a battery of tests. Let us give two examples.

A.1   Identity

The principle $A \vdash A$, the identity "axiom", poses a problem of finiteness. It is tested through simultaneous tests, for $\sim A$ and A, which is possible in certain cases, but doesn't work in general.

Let us suppose that A correspond to general behaviour **A**, with not finiteness restriction. I still know how to justify $\vdash \sim A, A$ because it is *sufficient* to test it against generic pairs, that of a test for $\sim \mathbf{A}$ and for **A** with no reference to **A** which therefore takes the moral value of a variable X. We know that the cases:

$$A = \daleth \qquad \sim A = \daleth \tag{23}$$

$$A = \daleth \otimes \daleth \qquad \sim A = \daleth \,\gimel\, \daleth \qquad (24)$$

$$A = \daleth \,\gimel\, \daleth \qquad \sim A = \daleth \otimes \daleth \qquad (25)$$

do suffice. They force the presence of a star $[\![\sim A(x), A(x)]\!]$, if I abusively denote the respective locations of $\sim A$ and A by $\sim A(x), A(x)$. This implies in turn that the said star does belong to the behaviour $\vdash \sim A, A$.

These tests are not necessary: if $A = B \otimes C$ and $\vdash \sim A, A$ has been obtained by "η-expansion" from $\vdash \sim B, B$ and $\vdash \sim C, C$, they fail.

We just witnessed the native *sufficient* testing. Remark that its two parts are not independent: if A is tested as $\daleth \otimes \daleth$, $\sim A$ must be tested as $\daleth \,\gimel\, \daleth$.

A.2 Existence

Existence can be informally reduced to a very peculiar case, that of the implication $\forall X\, A \vdash A[T/X]$, in other terms $\vdash \exists X \sim A, A[T/X]$. We must test $(\mathcal{T}, \mathcal{T}', \mathcal{P})$ where $\mathcal{P}$ is the identity $\vdash \sim A[T/X], A[T/X]$. We just observed that this identity possesses a sufficient battery of tests. We conclude that $\mathcal{P}$ belongs to the behaviour associated with $\vdash \sim A[T/X], A[T/X]$.

In order to show that $(\mathcal{T}, \mathcal{T}', \mathcal{P})$ is in the behaviour (7) corresponding to $\vdash \exists X \sim A, A[T/X]$, we imitate the argument given for system $\mathbb{F}$: the comprehension principle shows that the behaviour associated with T is a set, and we use the "substitution lemma" of [3].

A.3 Finitism

The finitistic pattern advocated by Hilbert is correct provided we throw in some necessary distinctions. Three layers are needed:

Usine: Typing à la Church, but system-free. It enables us to *predict* the behaviour are doing.

Usage: Typing à la Curry, naturally system-free. A *behavourial* approach, what proofs are actually doing.

Adequation: Cut-elimination, so to speak. It shows the accuracy of the prediction of l'usine.

The first two layers are the opposite sides of finitism, of a completely different nature. The first person who (vaguely) understood the distinction was Lewis Carroll (1893), who mistook l'usine for the "meta" of l'usage and built a ludicrous wild goose chase which he dared compare with Zeno's paradox. Indeed, by replacing a cut on A with a cut on $A \Rightarrow A$, next a cut on $(A \Rightarrow A) \Rightarrow (A \Rightarrow A)$, etc. Carroll's Achilles is constantly fleeing away from the Tortoise... no wonder it never reaches him.

The third layer, adequation, does not belong to finitism: it is where an infinitary, eventually set-theoretic, argument must be thrown in... with no possible way of bypassing it.

VITAM IMPENDERE LOGICÆ

References

1. Coquand, T., Huet, G.: The calculus of constructions. Inf. Comput. **76**, 95–120 (1988)
2. Danos, V., Regnier, L.: The structure of multiplicatives. Arch. Math. Logic **28**, 181–203 (1989)
3. Girard, J.-Y.: Une extension de l'interprétation fonctionnelle de Gödel à l'analyse et son application à l'élimination des coupures dans l'analyse et la théorie des types. In: Fenstad (ed.) Proceedings of the 2nd Scandinavian Logic Symposium, pp. 63–92. North-Holland, Amsterdam (1971)
4. Girard, J.-Y.: Transcendental syntax 1: deterministic case. Mathematical Structures in Computer Science, pp. 1–23 (2015). Computing with lambda-terms. A special issue dedicated to Corrado Böhm for his 90th birthday
5. Girard, J.-Y.: Transcendental syntax 3: equality. Logical Methods in Computer Science (2016). Special issue dedicated to Pierre-Louis Curien for his 60th birthday
6. Girard, J.-Y., Scedrov, A., Scott, P.: Bounded linear logic: a modular approach to polynomial time computability. Theor. Comput. Sci. **97**, 1–66 (1992)
7. Harper, R., Honsell, F., Plotkin, G.: A framework for defining logics. LFCS report series, Edinburgh, 162 (1991)
8. Kreisel, G.: Mathematical logic. In: Saaty, T.L. (ed.) Lectures in Modern Mathematics, vol III, pp. 99–105. Wiley, New York (1965)

Logic and Computing

A Small Remark on Hilbert's Finitist View of Divisibility and Kanovich-Okada-Scedrov's Logical Analysis of Real-Time Systems

Mitsuhiro Okada[⊠]

Keio University, 2-15-45 Mita, Minato-ku, Tokyo, Japan
mitsu@abelard.flet.keio.ac.jp

Abstract. Hilbert remarked in the introductory part of his most famous finitism address (1925 [1]) that "[t]he infinite divisibility of a continuum is an operation that is present only in our thought", which means that no natural event or matter is infinitely divisible in reality. We recall that Scedrov's group including the author started logical analysis of real time systems with the principle similar to Hilbert's no-infinite divisibility claim, in [2]. The author would like to note some early history of the group's work on logical analysis of real time system as well as some remark related to Hilbert's claim of no-infinite divisibility.

Keywords: Real-time system · Hilbert · Finitism · Zeno paradox · Andre Scedrov · Multi-agent system · Formal verification

It was at a narrow corridor of the faculty building that a big whiteboard (with a photo-copying function of the board) was temporarily placed, and the three researchers discussed for the whole days and formed a starting point of the logical analysis of real-time systems. Since it was necessary for all faculty members on the floor to pass along the corridor quite frequently everyday, they were required to get sophisticated skills to find a way to go through some narrow space among the big board and the three men discussing including the big guy with big voice. It was how the collaboration began among the three lead by Andre Scedrov at Keio University in Tokyo.

1 Introduction

In his most famous finitism address, which was published in [1], Hilbert claimed that infinite divisibility of natural events and mater was not real but just our naïve impression; in the other words, infinite divisibility of a continuity was "an operation that is present only in our thought" and "merely an idea", which is not in reality.

V. Nigam et al. (Eds.): Scedrov Festschrift, LNCS 12300, pp. 39–47, 2020.
https://doi.org/10.1007/978-3-030-62077-6_3

"The initial, naïve impression that we have of natural events (Geschehen) and of matter is one of uniformity, of continuity. If we have a piece of metal or a volume of liquid, the idea impresses itself upon us that it is divisible without limit, that any part of it, however small, would gain the same properties. But, wherever the methods of the research in the physics of matter were refined sufficiently, limits to divisibility were reached that are not due to the inadequacy of our experiments but to the nature of the subject matter...."

He continued, after mentioning the cases of electricity and quantum of energy, as follows.

"The infinite divisibility of a continuum is an operation that is present only in our thought; it is merely an idea, which is refused by our observations of nature, and by the experience gained in physics and chemistry (Hilbert 1925 [1])."

Here, Hilbert expresses an aspect of the finitist distinction between reality and ideas as the distinction between no-infinite divisibility in reality and infinite divisibility only in our thought, in the introductory part of this paper on his finitist program. His finitist program was, very roughly speaking, a safety verification program, to prove consistency of ideal reasoning of classical mathematics by means of finitist methods, in other words, by means of concrete and contentual methods, so that the freedom of the use of classical mathematical reasoning and of its applications to physics and other mathematically described sciences were provably-safe without any possibility of critical state "contradiction" (even though classical mathematical reasoning using continuity, infinite divisibility, set-theoretical infinity, etc.). Although at the above quoted part Hilbert talked about no-infinite divisibility of natural events and matter in the spatial continuity sense, one could presume, from the finitist view, the same claim of no-infinite divisibility for events in wider and various senses) in a spatiotemporal continuum, e.g., to claim no-infinite temporal divisibility of any, either natural or artificial, events.

Hilbert himself made the remark on no-infinite divisibility in reality from the finitist view in the context of the foundational studies in mathematics in the early 20th century. We would like to remark that the similar no-infinite divisibility settings have appeared in more contemporary contexts of computer science. In this Note, we recall one example investigated by Andre Scedrov's group including the author, initiated by [2], where some logical analyses of multi-agent real-time systems with dense-time were given in a logical way (using a subsystem of linear logic, or called multiset rewriting system). A different approach was given in using the timed automata setting.[1] This Note does not include the precise logical definitions and does not require the reader to know them, but the reader who wish to know more precise settings is invited to see the papers in the References.

In order to see the effect of the no-infinite divisibility setting, which we call the no-infinite divisibility principle in this Note, we could compare two different ways

---

[1] We only consider the logical approach in this Note.

of resolving Zeno's paradoxes, such as the Paradox of Achilles and the Tortoise. One way is the usual way, namely the classical analytic way, and the other way is the way employing the no-divisibility principle (without analytic notion of "limit"). In the Hilbert's finitist terminology, the one by using ideal operations of thought, the other by finitistic description of the real world. First, we remind ourselves what was Zeno's argument: The starting point of the Tortoise is a little ahead of that of Achilles. Zeno argues that when one considers the event that Achilles reaches the starting point of the Tortoise, the Tortoise's position is already a little advanced. Then, one considers another event that Achilles reaches the position of the Tortoise at the previous event, and in the same way one could consider unlimited number of events, which are all sub-events of this whole race event. Zeno claims that since there are infinitely many events before Achilles reaches the Tortoise, Achilles would never catch up with the Tortoise, which is a paradox.

However, by means of the modern mathematical definition of "limit" and convergence of an infinite series, one can easily calculate exactly when and where in the spatiotemporal continuum Achilles catches up with the Tortoise and overtakes hime. This usual resolution is consistent with Zeno's argument itself, without any contradiction assuming that modern mathematics has no contradiction.[2]

The other way to resolve the Paradox of Achilles and the Tortoise is to take the standpoint of the no-divisibility principle that there is no infinite divisibility of the race: Zeno's argument divides the race into infinite sub-events, which violates reality, namely the no-infinite divisibility principle. Hence, the principle blocks Zeno's argument itself, therefore no paradox occurs here. From the view point of Hilbert, the former analytic resolution is not real but ideal, while the latter real. This latter would suggest us some idea of finite states and events-based modeling of the outer world of finitely many agents on the spatiotemporal continuum with the no-infinite divisibility principle. For example, the race of Achilles and the Tortoise could be modeled as a two-agent state transition system in various ways by choosing primitive states and by respecting the no-infinite divisibility principle. An extremely simplest modeling, but with reality, is the following.

In "What the Tortoise Said to Achilles" (1895), Lewis Carroll emphasizes that reality of the race uses only a few state transitions; with starting state of the race, only after a single tick (a little time progress, expressed by a single application of the tick-axiom in the sense of [2], it gets the state that Achilles has overtaken the Tortoise, then, after another-tick, Achilles has seated himself on the back of the Tortoise. There would be various different ways to set the primitive states associated to each agent to describe the race or other events in a state transition-based multi-agent real-time systems depending on a focused critical states or events.[3] One could consider more complexed multi-agent systems

---

[2] To show no contradiction of classical mathematical reasoning was the exact purpose of Hilbert's finitist program.

[3] Note that Zeno, believing his argument of the paradox successful, concluded no state-change and no event in reality of the world, which is too simple description of the real world which is contradictory with our intuition.

where various time constraints are imposed on the time-continuum. For example, if we have one-time constraint that the race should be finished by 5 h after the starting-time (as such a time constraint is usually imposed to some citizen marathon events), then, Achilles who now sits on the back of the Tortoise automatically moving towards the goal, needs to think a scheduling until when he may remain on the back of the Tortoise and when he should leave the back to start running himself) (or just one jump) to the goal in the constraint deadline, namely safely with respect to the critical state. Adding time constraints would involve notions of "may", "should", "critical state", "safety", "scheduling" in multi-agent modeling with the no-infinite divisibility principle even for modeling the Achilles-the Tortoise race by a state transition system.

2   A Case Study: Multi-agent Real-Time State Transition Model with the No-Infinite Divisibility Principle

2.1   Finitely-multi-agent Finite State Transition System with Finite Time Constraint Conditions on the Dense Real-Time

Kanovich-Okada-Scedrove [2] provided a case study of finitely modeling of real time system with the no-infinite divisibility principle, which is sometime also called the Zeno principle, which tells that there is no infinite events, actions, or state transitions in any bounded time interval of the dense or continuous time.[4] Kanovich-Okada-Scedrov [2] took a state transition model of the external environments and any state is either a primitive state or a composite state of concurrent finite primitive states. Any primitive state is typically related to an agent. A composite state transition was expressed by a logical implication where a composite state is expressed by multiplicative connectives of atomic states, which has been also called a multiset rewrite rule by Scedrov's group later (See References, where the reader could find the use of the name even in the title of later papers co-authored by Scedrov.). A time progress with unspecified duration is expressed by a time transition rule called tick-axiom. For example, instead of (classical) mathematical description of physical moving as to how a train approaches to a railway gate by means of continuous function on the real number, one takes a finite-states description, such as "far distance safe state", "approaching to the railway (be-careful) state", "train's passing the railway gate (danger) state", "train's-passing away state" for the purpose of safety verification of a traffic system as in [2]. The distinction of the two ways of description of train states has a certain correspondence to the distinction between the classical mathematical-physical description with continuity and the finite description with the no-infinite divisibility of events, discussed in the previous Section. Scedrov's group initiated by [2] showed that the apparent dense real-time is eliminable; the main purpose was to specify a certain class of dense-time multi-agent systems

---

[4] Setting a time interval could be considered an event with the duration, such as the race of Achiles and the Tortoise.

with time-constraints, with the focuses on (1) showing formal specification ability of critical states for safety and (2) showing computational complexity of the reachability problem, which implies the complexity of decision problem of safety in the specification framework. The dense-time setting provided an interesting modeling of multi-agent system, where any agent could start its primitive state transition (or an agent's action) at any moment on the dense time under finitely many must-time constraints and may-time constraints.[5] This dense-time setting could allow us to have an interpretation that any participating agent apparently has freedom to trigger its action; e.g., a walker can change her state from the standing-state to the crossing-street-state whenever some may-constraints and must-constraints are satisfied; a may-constraint could be "one may start crossing 5 s after the traffic-light changed to the Green state" , and a must-constraint could be "one must complete crossing the street before 5 s passing of the Yellow state of the traffic light", for example. [2] and the related works tell that in fact the dense time is eliminable with the state transition model and the reachability problem is PSPACE. This could be understood that if a given specific composite state happens or not is PSPACE decidable even if apparent free-triggering primitive state transitions by any agent under the constraints. There are some corresponding work in the timed-automata research community.

2.2 Global Clock and Local Alarm Stopwatches

The original work [2] was involved with set theoretical treatment to consider equivalent classes of states to eliminate the dense (infinite) time-states. During the author's stay in Paris in 2000, the author had an opportunity to reformulate it by switching the use of global clocks to local alarm-stopwatches and gave lectures on it. The author shared this modified method with the co-authors of [2]. This switching gave the effect of putting the time constraint locally inside the linear logical implicational or multiset transition rules and the whole argument of [2] became more constrictive in the sense that one eliminates the dense-time without the set-theoretical equivalence classes treatment[6].

To hint some of the specific characteristics of the work initiated by [2] it would be worth noting here various different reactions from the audience to the author's series of lectures on this topic in 2000–2002. For example, when the author gave an invited talk at an interdisciplinary meeting organized by G. Longo at ENS-Paris, it was physicists who expressed most appreciations on our work, especially on the way how the continuous time in the physics sense could be reformed to finite state transition systems with PSPACE decidability. As for another example, when the author gave the lectures at the Coq group meetings in INRIA, they were interested in the way to switch the global clock to local

---

[5] A transition of primitive state associated to an agent is called an action of the agent in this Note for simplicity of the argument. See [2] for more formal and precise setting.

[6] Among our agreement Max Kanovich was supposed to write the version, but unfortunately without realization soon.

watches to make the argument more constructively, and the author collaborated with them on the topic including an implementation work. When the author gave talks to the audience of the AI and multi-agent system research community, some of them told us that it was such a rare good work in the AI-multi-agent system research community that the primitive concepts, ontology, syntax, technical proofs of computational complexity, semantics, applications and implementation altogether were given by one group: they appreciated our work. The part of our local-watches-based linear logical transition systems were implemented by our master students on the Coq group under the author's supervision with Jean-Pierre Jouannaud, as FATALIS System. Some part of FATALIS was presented in [4]. Some versions were presented at [3] and [5] by the author. The linear logical (multiset reriwrite) system framework is easy to express dynamic changes of the system, such as new agents participations, changing the time constraints, on the rewrite rule level.

The framework proposed in [2] was one of the earliest works for modeling timed multi-agent systems, either with dense-time or with discrete time, which opened a wide range of applications. Andre Scedrov and his colleagues have developed sophisticated logical techniques on this line for various different settings, including time-related security, scheduling and others. See [6–9] for some of the recent developments of the group. These show wide range of describability and logical analyzability by means of the timed multiset rewriting framework and its further potentials.

2.3   Back to Finitism

We already remarked that the no-infinite divisibility claim, in Hilbert's finitism [1], also occured in the setting of [2] as the basic principle. Hence it would be interesting to comparing the two shortly to conclude this Note. Hilbert distinguishes reality and ideality, and he anticipated a finitist or realist (in his sense) safety verification proof of classical or ideal (in his sense) mathematics which was essentially needed for our describing the physical world. On the other hand, the standpoint of the framework of the timed multiset rewriting (and its linear logical version) is rather neutral with respect to the reality-vs-ideality distinction of the description level. It could be considered a combined description. For example, the timeless version of the discrete system description of [2] is a finitely modeling (hence idealized modeling) for some focused aspect of the external world, such as a part of train-traffic system for the purpose of safety proofs or safe scheduling etc. But, when the dense time enters to the discrete model, the no-infinite divisibility is required as a realist requirement. (As we reminded in Introduction, unless one takes the continuous physical description of the agents' moving, it would go into Zeno's too idealistic world view.) One could even relativize the real-time; in certain finitely described cases discussed in this Note the no-infinite divisibility is the real requirement, while in certain analytically-physically described cases the continuity is the real requirement. The finitely described real-time modeling framework used in [2] and others provided typical cases of the former. A discrete state transition modeling could be natural as a modeling of the physical real train traffic in "reality". In fact, when we define which area is the "approaching (danger) state" area, for example, we need some physical calculation; then based on the physical calculation, one takes the view of train traffic system as a discrete state transition system, In that case, one might think that the description by continuous functions with the time-continuum as relatively real and the description by discrete modeling relatively with the no-infinite divisibility relatively ideal. However, one could easily imagine the other way around, too, for cetain cases. After designing and formally verifying safety with the timed multiset rewrite system, the physical area and the physical speed bound of the approaching-state can be defined; in this way, one could consider the descrete safe system is relatively real and the implementation in the physical world is a simulation of the symbolic safe system. In fact, both views are interactive. Time-continuum and no-infinite divisibility of time co-exist. One could say the interactive view of the two aspects important, beyond the reality-indeality issue. In practice, we often use the view of hybrid systems accommodating both.

Note that we have not discussed another way of description, besides the finite-descrete way and the analytic mathematical -physical way of description (say, for safety verification): one missing was the probabilistic way of description. For example, security properties of cryptographic protocols are often investigated and proven by the discrete modeling, such as using finitely Dolev-Yao model description, which can be understood as a state transition model, and by the cryptographic computational description, such as using probabilistic poly-time Turing machine description and others. The interactive view of the two ways of description is important in these cases, too. One of the author's views may be found in [12] where a variant of the forcing model construction, Fitting-embedding, method was used for connecting the logical-descrete verification proofs to the poly-time probabilistic Turing machine based-verification proofs, by means of giving "computational soundness" of logical proofs with respect to the computational models.[7]

Acknowledgements. This work was supported by JSPS KAKENHI Grant Numbers 19KK0006, 17H02265, 17H02263. The author would like to express his sincere thanks to the anonymous referee for helpful comments at the preparation stage of this manuscript.

---

[7] Andre Scedrov and his colleagues are known as the experts on the domain of formal verification of time-related cryptographic protocols using the similar framework discussed in this Note. The author have been working on clarifying the relationship between the two ways of description for security verifications, including [12], especially with Gergei Bana, a former student of Andre Scedrov.

References

1. Hilbert, D.: On the infinite, in Frege to Gödel, pp. 369–392. Harvard University Press (1967). (Edited By van Heijenoort, J.) The original version: Über das Unentliche, Mathematische Annalen **95**, 161–190 (1926). Jahresbericht der Deutschen Mathematiker-Vereinigung, **36**, 201–215 (1927)
2. Kanovich, M.I., Scedrov, A., Okada, M.: Specifying real-time finite-state systems in linear logic. In: 2-nd International Workshop on Constraint Programming for Time-Critical Applications and Multi-agent Systems (COTIC), Electronic Notes in Theoretical Computer Science, Nice, France, vol. 16, no. 1, September 1998 (1998). 15 pp
3. Okada, M.: Theory of formal specification and verification of concurrency systems and real-time systems based on linear logic, Report Meeting, Jan. 2002 Mext 12480075, also, Logical verification method for dynamic real-time system beyond the limit of model-checking method, Report Meeting, MEXT: 13224081
4. Hasebe, K., Cremet, V., Jouannaud, J.-P., Kremer, A., Okada, M.: FATALIS: real time processes as linear logic specifications. In: Second International Workshop on Automated Verification of Infinite-State Systems (2003)
5. Hasebe, K., Jouannaud, J.-P., Kremer, A., Okada, M., Zumkeller, R.: Formal verification of dynamic real-time state-transition systems using linear logic. In: The 20th Conference of Software Science Society of Japan, 2003, Proc., Japan (2003)
6. Alturki, M.A., Ban Kirigin, T., Kanovich, M., Nigam, V., Scedrov, A., Talcott, C.: A multiset rewriting model for specifying and verifying timing aspects of security protocols. In: Guttman, J.D., Landwehr, C.E., Meseguer, J., Pavlovic, D. (eds.) Foundations of Security, Protocols, and Equational Reasoning. LNCS, vol. 11565, pp. 192–213. Springer, Cham (2019). https://doi.org/10.1007/978-3-030-19052-1_13
7. Kanovich, M.I., Ban Kirigin, T., Nigam, V., Scedrov, A., Talcott, C.L.: Compliance in Real Time Multiset Rewriting Models. CoRR abs/1811.04826 (2018)
8. Kanovich, M.I., Ban Kirigin, T., Nigam, V., Scedrov, A., Talcott, C.L., Ranko, P.: A rewriting framework and logic for activities subject to regulations. Math. Struct. Comput. Sci. **27**(3), 332–375 (2017)
9. Kanovich, M.I., Ban Kirigin, T., Nigam, V., Scedrov, A., Talcott, C.L.:Timed multiset rewriting and the verification of time-sensitive distributed systems. In: FORMATS, pp. 228–244 (2016)
10. Kanovich, M.I., Ban Kirigin, T., Nigam, V., Scedrov, A., Talcott, C.L.: Discrete vs. dense times in the analysis of cyber-physical security protocols. In: POST, pp. 259–279 (2015)
11. Kanovich, M.I., Ban Kirigin, T., Nigam, V., Scedrov, A., Talcott, C.L., Ranko, P.: A Rewriting Framework for Activities Subject to Regulations. In: RTA, pp. 305–322 (2012)
12. Bana, G., Okada, M.: Semantics for "Enough-Certainty" and Fitting's embedding of classical logic in S4. In: Computer Science Logic 2016, Proceedings, article 23, 17 p. (2016)

Logic of Fusion

— *Dedicated to Andre Scedrov* —

Dusko Pavlovic[✉]

University of Hawaii, Honolulu, HI, USA
dusko@hawaii.edu

Abstract. We pursue an extension of the Curry-Howard isomorphism of propositions and types by a correspondence of cut elimination and program fusion. In particular, we explore the repercussions of this extension in generic and transformational programming. It provides a logical interpretation of build fusion, or deforestation, in terms of the inductive and the coinductive datatypes. Viewed categorically, this interpretation leads to the novel structure of paranatural transformations. This is a modified version of functorial polymorphism, that played a prominent role in the work of Andre Scedrov.

Personal Introduction

I first met Andre at one of the Peripatetic Seminars on Sheaves and Logic (PSSL) in the late 80s. PSSL was a legendary community of category theorists, meeting a couple of times a year at venerable universities in Europe and the UK. Andre was a well-established researcher, who had already subsumed forcing under the classifying topos construction; and I was a wide-eyed grad student. He was pointed out to me as coming from the same country where I had come from (which at the time still existed); but the main reason why I had already read maybe not all, but most of his papers, was that I was trying to learn and understand the powerful new methods of category theory that Andre had worked on.

Nowadays, you probably wouldn't call either Andre or me a category theorist. The word "category" does not occur that often either in his or in my papers. Yet, if you follow the common thread that ties together Andre's work, it takes you through logic, semantics of computation, decision procedures and algorithms, models of natural language, security protocols. It is a very long thread. Longer than just a thread of good taste, of clever constructions, and honest excitement. It is a thread of method. By trying to trace this thread through Andre's work, I confront the challenge of explaining my own. How did we get from toposes and constructivist universes to distance bounding protocols and mafia attacks? Of course I don't know the answer. One answer might be that we got wiser. Another answer is that we are still too young to tell. Or is that just a wise way to avoid answering?

D. Pavlovic—Supported by NSF and AFOSR.

V. Nigam et al. (Eds.): Scedrov Festschrift, LNCS 12300, pp. 48–60, 2020.
https://doi.org/10.1007/978-3-030-62077-6_4

Instead of attempting to answer, or to avoid answering, I recall an intermediary step. I reproduce for the occasion a construction from a long time ago, that arose when I read [14], and then [2, 4, 8]. The construction was never published, although it indirectly led to [11]. It was developed for specific applications in a tool that I was trying to build [1], but the conceptual problem was reduced to the toy task of polymorphic zipping. By that time, Andre was already past the polymorphism research phase. When I caught up with him at the next corner, we were both thinking about security. There seems to be some sort of polymorphism behind it all.

1   Technical Introduction

1.1   Idea

The starting point of this work is the observation that the Curry-Howard isomorphism [16], relating

$$\text{types} \leftrightsquigarrow \text{propositions}$$
$$\text{programs} \leftrightsquigarrow \text{proofs}$$
$$\text{composition} \leftrightsquigarrow \text{cut}$$

can be extended by a correspondence of

program fusion $\leftrightsquigarrow$ cut elimination

This simple idea suggests logical interpretations of some of the basic methods of generic and transformational programming. In the present paper, we provide a logical analysis of the general form of *build fusion*, also known as *deforestation*, over the inductive and the coinductive datatypes, regular or nested. The analysis is based on a logical reinterpretation of parametricity [17] in terms of *paranatural* transformations, modifying the functorial interpretation of polymorphism in [2].

1.2   Fusion and Cut

The Curry-Howard isomorphism is one of the conceptual building blocks of type theory, built deep into the foundation of computer science and functional programming [10, Ch. 3]. The fact that it is an *isomorphism* means that the type and the term constructors on one side obey the same laws as the logical connectives, and the logical derivation ruleson the other side. For instance, the products and the sums of types correspond, respectively, to the conjunction and the disjunction, because the respective introduction rules

$$\frac{A \vdash B_0 \quad A \vdash B_1}{A \vdash B_0 \wedge B_1} \wedge I \qquad\qquad \frac{A_0 \vdash B \quad A_1 \vdash B}{A_0 \vee A_1 \vdash B} \vee I$$

extended by the labels for proofs, yield the type formation rules

$$\frac{f_0 : A \to B_0 \quad f_1 : A \to B_1}{\langle f_0, f_1 \rangle : A \to B_0 \times B_1} \qquad \frac{g_0 : A_0 \to B \quad g_1 : A_1 \to B}{[g_0, g_1] : A_0 + A_1 \to B}$$

In a sense, the pairing constructors $\langle -, - \rangle$ and $[-, -]$ record on the terms the applications of the rules $\wedge I$ and $\vee I$, as the proof constructors.

Extending this line of thought a step further, one notices that the term reductions also mirror the proof transformations. E.g., the transformation

$$\frac{\dfrac{A_0 \vdash B \quad A_1 \vdash B}{A_0 \vee A_1 \vdash B} \quad B \vdash C}{A_0 \vee A_1 \vdash C} \quad \blacktriangleright\blacktriangleright \quad \frac{\dfrac{A_0 \vdash B \quad B \vdash C}{A_0 \vdash C} \quad \dfrac{A_1 \vdash B \quad B \vdash C}{A_1 \vdash C}}{A_0 \vee A_1 \vdash C}$$

corresponds to the rewrite

$$h \cdot [f_0, f_1] \quad \blacktriangleright\blacktriangleright \quad [h \cdot f_0 , \; h \cdot f_1] \tag{1}$$

where f_0 and f_1 are the labels of the proofs $A_0 \vdash B$ and $A_1 \vdash B$, whereas h is the label of $B \vdash C$. The point of such transformations is that the applications of the cut rule

$$\frac{A \vdash B \quad B \vdash C}{A \vdash C} \tag{2}$$

get pushed up the proof tree, as to be eliminated, by iterating such moves. On the side of terms and programs, the cut, of course, corresponds to the composition

$$\frac{f : A \to B \quad h : B \to C}{h \cdot f : A \to C} \tag{3}$$

Just like the presence of a cut in a proof means that an intermediary proposition has been created, and then cut out, the presence of the composition in a program means that the thread of computation leads through an intermediary type, used to pass data between the components, and then discarded. Computational aspects of normalization are discussed in [10, Ch. 4].

While the programs decomposed into simple parts are easier to write and understand, passing the data and control between the components incurs a computational overhead. For instance, running the composite ssum · zipW of

```
        zipW                : [Nat]×[Nat] -> [Nat×Nat]
        zipW (x::xs,y::ys) = (x,y) :: zip xs ys
        zipW (xs,    ys)   = []
```

and

```
        ssum                : [Nat×Nat] -> Nat
        ssum []            = 0
        ssum (x,y)::zs = x + y + sum zs
```

is clearly less efficient than running the fusion

```
sumzip                  : [Nat]×[Nat] -> Nat
sumzip (x::xs,y::ys)  = x + y + sumzip (xs,ys)
sumzip (xs,   ys)     = 0
```

where the intermediary lists [Nat × Nat] are eliminated. In practice, the data structures passed between the components tend to be very large, and the gain by eliminating them can be significant. On the other hand, the efficient, monolythic code, obtained by fusion, tends to be more complex, and thus harder to understand and maintain.

To get both efficiency and compositionality, to allow the programmers to write simple, modular code, and optimize it in compilation, the program fusions need to be sufficiently well understood to be automated. Our first point is that the Curry-Howard isomorphism maps this task onto the well ploughed ground of logic.

1.3 Build Fusion

The general form of the build fusion that we shall study corresponds, in the inductive case, to the "cut rule"

$$
\frac{A \xrightarrow{f} M_F \qquad
\begin{array}{ccc}
FM_F & \xrightarrow{F(\!|c|\!)} & FC \\
{\scriptstyle\mu}\downarrow & & \downarrow{\scriptstyle c} \\
M_F & \xrightarrow{(\!|c|\!)} & C
\end{array}}
{A \xrightarrow{f'C(\ulcorner c\urcorner)} C}
\tag{4}
$$

eliminating the inductive datatype M_F, which is the initial algebra of the type constructor F. In practice and in literature, F is usually a list- or a tree-like constructor, and the type A is often inductively defined itself; but we shall see that the above scheme is valid in its full generality. The sumzip-example from the preceding section can be obtained as an instance of this scheme, taking $FX = 1 + \mathtt{Nat} \times \mathtt{Nat} \times X$, and thus $M_F = [\mathtt{Nat} \times \mathtt{Nat}]$. The function ssum is the catamorphism (fold) of the map $[0, \ddagger] : 1 + \mathtt{Nat} \times \mathtt{Nat} \times \mathtt{Nat} \longrightarrow \mathtt{Nat}$ where $\ddagger$ maps $\langle i, j, k \rangle$ to $i + j + k$.

The dual scheme

$$
\frac{
\begin{array}{ccc}
FA & \xrightarrow{F[\!(a)\!]} & FN_F \\
{\scriptstyle a}\uparrow & & \uparrow{\scriptstyle \nu} \\
A & \xrightarrow{[\!(a)\!]} & N_F
\end{array}
\qquad N_F \xrightarrow{g} C}
{A \xrightarrow{g'A(\ulcorner a\urcorner)} C}
\tag{5}
$$

allows eliminating the coinductive type N_F, the final F-coalgebra.

Clearly, the essence of both of the above fusion schemes lies in the terms f' and g'. Where do they come from? The idea is to represent the fixpoints M_F and N_F in their "logical form"

$$M_F \cong \forall X. \ (FX \Rightarrow X) \Rightarrow X \qquad (6)$$
$$N_F \cong \exists X. \ X \times (X \Rightarrow FX) \qquad (7)$$

The parametric families

$$f'X : (FX \Rightarrow X) \longrightarrow (A \Rightarrow X) \qquad (8)$$
$$g'X : (X \Rightarrow FX) \longrightarrow (X \Rightarrow C) \qquad (9)$$

are then obtained by extending $f : A \longrightarrow M_F$ and $g : N_F \longrightarrow C$ along isomorphisms (6) and (7), and rearranging the arguments. The equations

$$(|c|) \cdot f = f'C(\ulcorner c \urcorner) \qquad (10)$$
$$g \cdot (\!|a|\!) = g'A(\ulcorner a \urcorner) \qquad (11)$$

can be proved using logical relations, or their convenient derivative, Wadler's "theorems for free" [18]. This was indeed done already in [9] for (10), and (11) presents no problems either.

However, mapped along the Curry-Howard isomorphism, Eqs. (10–11) become statements about the equivalence of proofs. The fact that all logical relations on all Henkin models must relate the terms involved in these equations does not seem to offer a clue for understanding their equivalence.

Overview of the paper

In order to acquire some insight into the logical grounds of program fusion, and equivalence, we propose *paranatural* transformations, presented in Sect. 2, as a conceptually justified and technically useful instance of the dinatural semantics of polymorphism [2]. The applicability of this concept is based on the characterization of the parametricity of families (8) and (9) in terms of an intrinsic commutativity property. We note that this characterization is completely intrinsic, with no recourse to models or external structures. The upshot is that the results actually apply much more widely than presented here, i.e. beyond the scope of build fusion. But that was the application that motivated the approach, and it suffices to show the case. The paranaturality condition is a variation on the theme of functorial and structural polymorphism [2,4,7,8,14]. Unfortunately, neither of these semantical frameworks provides sufficient guidance for actual programming applications. The dinatural transformations of [2,8] provide conceptually clear view of polymorphism as an invariance property; but it has been recognized early on that the characterization is too weak, as it allows too many terms. On the other hand, the structor morphisms of [7] precisely correspond to the accepted polymorphic terms; but the approach is not effective, as it does not stipulate which of the many possible choices of structors should be used to interpret a particular polytype. We propose paranatural transformations as means for filling this gap. This proposal emerged from practical applications in programming. It is based on the insight, on the logical background of Propositions 1 and 2, that program fusion only ever requires capturing as polymorphic one of two kinds of families of computations:

- those where the inputs *from* some final datatypes are consumed, and
- those where the outputs are produced *into* some initial datatypes.

Proposition 3 in Sect. 3 formalizes this idea. The proof of this proposition is given in the Appendix. The proofs of the other propositions are straightforward. We note that the result eliminates the extensionality and the well-pointedness requirements of logical relations, which hamper their applications, even on the toy examples discussed here. On the other hand, refining the logical approach from Sects. 1.2 and 1.3 along the lines of [13] seems to broaden the presented methods beyond their current scope. Some evidence of this is discussed in the final section.

2   Paranatural Transformations

As it has been well known at least since Freyd's work on recursive types in algebraically compact categories [6], separating the covariant and the contravariant occurrences of X in a polytype $T(X)$ yields a polynomial functor $T : \mathbb{C}^{op} \times \mathbb{C} \longrightarrow \mathbb{C}$. On the other hand, by simple structural induction, one easily proves that

Proposition 1. *For every polynomial functor $T : \mathbb{C}^{op} \times \mathbb{C} \longrightarrow \mathbb{C}$ over a cartesian closed category $\mathbb{C}$, there are polynomial functors $W : \mathbb{C}^{op} \times \mathbb{C} \longrightarrow \mathbb{C}$ and $V : \mathbb{C} \longrightarrow \mathbb{C}$, unique up to isomorphism, such that*

$$T \cong W \Rightarrow V$$

This motivates the following

Definition 1. *Let $\mathbb{C}$ be a category and $W : \mathbb{C}^{op} \times \mathbb{C} \longrightarrow C$ and $V : C \longrightarrow \mathbb{C}$ functors on it.*
A paranatural *transformation $\vartheta : W \longrightarrow V$ is a family of $\mathbb{C}$-arrows $\vartheta X : WXX \longrightarrow VX$, such that for every arrow $u : X \longrightarrow Y$ in $\mathbb{C}$, the external pentagon in the following diagram*

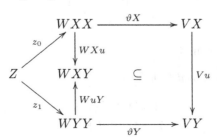

commutes whenever the triangle on the left commutes, for all Z, z_0 and z_1 in $\mathbb{C}$. This conditional commutativity is annotated by the $\subseteq$ inside the diagram. The class of the paranatural transformations from W to V is written $\mathsf{Para}(W, V)$.

Remark. When $\mathbb{C}$ supports calculus of relations, the quantication over Z, z_0 and z_1 and the entire triangle on the left can be omitted: the definition boils down to the requirement that the square commutes up to $\subseteq$, in the relational sense.

Proposition 2. *Let $\mathcal{L}$ be a polymorphic λ-calculus, and $\mathbb{C}_{\mathcal{L}}$ the cartesian closed category generated by its closed types and terms. For every type constructor T, definable in $\mathcal{L}$, there is a bijective correspondence*

$$\mathbb{C}_{\mathcal{L}}\left(A, \ \forall X.T(X)\right) \cong \mathsf{Para}(A \times W, V)$$

natural in A.

3   Characterizing Fixpoints

Proposition 3. *Let $\mathbb{C}$ be a cartesian closed category, and F a strong endofunctor on it. Whenever the initial F-algebra M_F, resp. the final F-coalgebra N_F exist, then the following correspondences hold*

$$\mathbb{C}(A, M_F) \cong \mathsf{Para}\left(A \times (FX \Rightarrow X), \ X\right) \tag{12}$$

$$\mathbb{C}(N_F, B) \cong \mathsf{Para}\left(X \times (X \Rightarrow FX), \ B\right) \tag{13}$$

naturally in A, resp. B.

The proof of this proposition is given in the Appendix.

   In well-pointed categories and strongly extensional λ-calculi, this propositoion boils down to the following "yoneda" lemmas.

Notation. Given $h : A \times B \longrightarrow C$ and $b : 1 \longrightarrow B$, we write $h(b)$ for the result of partially evaluating h on b

$$A \xrightarrow{\langle \mathrm{id}, b_! \rangle} A \times B$$

with arrow $h(b)$ from A to C and arrow h from $A \times B$ to C.

where $b_!$ denotes the composite $A \xrightarrow{!} 1 \xrightarrow{b} B$.

Lemma 1. *For paranatural transformations*

$$\varphi_X : A \times (FX \Rightarrow X) \longrightarrow X$$
$$\psi_Y : Y \times (Y \Rightarrow FY) \longrightarrow B$$

hold the equations

$$\varphi_X(\ulcorner x \urcorner) = (\!| x |\!) \cdot \varphi_{M_F}(\mu) \tag{14}$$
$$\psi_Y(\ulcorner y \urcorner) = \psi_{N_F}(\nu) \cdot [\![y]\!] \tag{15}$$

for all $x : FX \longrightarrow X$ *and* $y : Y \longrightarrow FY$.

While (14) follows from

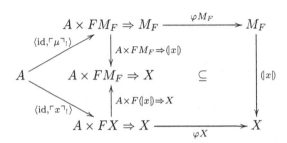

(15) is obtained by chasing

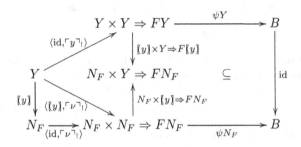

In well-pointed categories, $\varphi_X : A \times (FX \Rightarrow X) \longrightarrow X$ is completely determined by its values $\varphi_X(\ulcorner x \urcorner) : A \longrightarrow C$ on all $x : FX \longrightarrow X$. Similarly, $\psi_Y : Y \times (Y \Rightarrow FY) \longrightarrow B$ is completely determined by its values on $y : Y \longrightarrow FY$.

However, in order to show that $\varphi_{M_F}(\mu)$ is generic for φ and $\psi_{N_F}(\nu)$ for ψ without the well-pointedness assumption, one needs to set up slightly different constructions.

4   Applications

Using correspondence (12), i.e. the maps realizing it, we can now, first of all, provide the rational reconstruction of the simple fusion from the introduction. The abstract form of the function `zipW` will be

```
zipW' : [Nat]×[Nat] -> ((1+Nat×Nat×X)->X)->X
zipW' X (x::xs,y::ys) [m,c] = c(x, y, zipWith' X (xs,ys)
[m,c]) zipW' X (xs, ys) [m,c] = m
```

While `zipW` can be recovered as the instance `zipW'` [Nat × Nat] _ [[],(::)]`, i.e. `zipW = build(zipW')`, the fusion is obtained as

```
sumzip  =  zipW' Nat _ [0,‡]
```

But what is `zipW`, if it is not a catamorphism? How come that it still has a recursive definition?

It is in fact an *anamorphism*, and `ssum·zipW` can be simplified by the coinductive build fusion as well. The scheme is this time

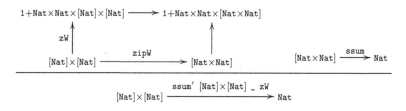

where

```
zW (x::xs,y::ys) = (x,y,xs,ys)
zW (xs,ys)       = One   (the element of 1)
```

induces $\texttt{zipW} = (\!|\texttt{zW}|\!)$, whereas

```
ssum'        : X × (X -> 1+Nat×Nat×X ) -> Nat
ssum' X x d = case d x of
                 One     -> 0
                 (n,m,y) -> n + m + ssum' X y d
```

Calculating the conclusion this time yields

```
sumzip = ssum' [Nat]×[Nat] _ zW
```

Finally, lifting proposition 3 to the category $\mathbb{C}^{\mathbb{C}}$ of endofunctors, we can derive the build fusion rule for nested data types [3]. Consider, e.g., the type constructor `Nest`, that can be defined as a fixpoint of the functor $\Psi : \mathbb{C}^{\mathbb{C}} \longrightarrow \mathbb{C}^{\mathbb{C}}$, mapping $\Psi(F) = \lambda X.1 + X \times F(X \times X)$.

The elements of the datatype `Nest Nat` are the lists where the i-th entry is an element of Nat^{2^i}. Abbreviating `Nest Nat` to {Nat}, we can now define

```
zWN (x::xs,y::ys) = (x,y,fst xs,fst ys,
                       snd xs,snd ys)
zWN (xs,ys)          = One
```

where `fst` and `snd` are the obvious projections $\{X \times X\} \longrightarrow \{X\}$, and and derive `zipWN : {Nat} × {Nat} ⟶ {Nat × Nat}` as $(\!|\texttt{zWN}|\!)$ again. On the other hand, working out the paranaturality condition in $\mathbb{C}^{\mathbb{C}}$ allows lifting

```
ssumN                : {Nat×Nat} -> Nat
ssumN []             = 0
ssumN (x,y)::zs = x + y + ssumN (fst zs)
                              + ssumN (snd zs)
```

to

```
ssumN'               : F(Nat) ×
                       F(X) -> 1+X×X×F(X×X) -> Nat
ssumN' F X f d = case d Nat f of
      One      -> 0
      (n,m,g) -> m + n + ssumN' FF X g dd
```

where **FF** and **dd** are the instances with $X \times X$ instead of **X**. The fusion

```
sumzipN = ssumN' Nest×Nest Nat _ zWN
```

is this time

```
sumzipN                  : {Nat}×{Nat} -> Nat
sumzipN (x::xs,y::ys)  = x + y + sumzipN (fst xs,fst ys) +
                                  sumzipN (snd xs,snd ys)
sumzipN (xs,   ys)     = 0
```

5 Afterword

The real application that motivated the presented work was a network application, based on event-channel architecture. A process involved a stream producer and a stream consumer, and the problem was to move filtering from the client side to the server side. Build fusion made this possible. The intermediary datatype, eliminated through build fusion, was thus infinitary: the streams. While the presented approach achieved its goal, and significantly improved the system, albeit in exchange for a lengthy derivation, the server at hand was actually a service aggregator, and thus also a client of other servers; and those wervers were for their part also other servers' clients. So there was a cascade of streams to be eliminated by means of a cascade of build fusions. The upshot is that the theoretical approach presented here simplified the practical application; but the practical application demonstrated that the calculations needed to apply the theory were intractably complex. The task of automating the approach opened up, and remained open. On the bright side, the event-channels involved security protocols. As I was trying to learn more about that, I realized that structural methods seemed to apply in that area as well, and that it was under active explorations by Andre Scedrov, with many friends and collaborators [5,12,15].

Appendix: Proof of Proposition 3

Towards isomorphism (12), we define the maps

$$(-)' : \mathbb{C}(A, M_F) \longrightarrow \mathsf{Para}\,(A \times (FX \Rightarrow X),\ X)$$
$$\mathtt{build} : \mathsf{Para}\,(A \times (FX \Rightarrow X),\ X) \longrightarrow \mathbb{C}(A, M_F)$$

and show that they are inverse to each other.

Given $f : A \longrightarrow M_F$, the X-th component of f' will be

$$f'_X \ : \ A \times (FX \Rightarrow X) \xrightarrow{f \times k} M_F \times (M_F \Rightarrow X)$$
$$\xrightarrow{\ \varepsilon\ } X$$

where $k : (FX \Rightarrow X) \longrightarrow (M_F \Rightarrow X)$ maps the algebra structures $x : FX \to X$ to the catamorphisms $(\!|x|\!) : M_F \to X$. Formally, k is obtained by transposing the catamorphism $(\!|\kappa|\!) : M_F \longrightarrow (FX \Rightarrow X) \Rightarrow X$ for the F-algebra κ on $(FX \Rightarrow X) \Rightarrow X$, obtained by transposing the composite

$$(FX \Rightarrow X) \times F\,((FX \Rightarrow X) \Rightarrow X) \longrightarrow$$
$$\xrightarrow{\text{(i)}} (FX \Rightarrow X) \times (FX \Rightarrow X) \times F\,((FX \Rightarrow X) \Rightarrow X)$$
$$\xrightarrow{\text{(ii)}} (FX \Rightarrow X) \times F\,((FX \Rightarrow X) \times (FX \Rightarrow X) \Rightarrow X)$$
$$\xrightarrow{\text{(iii)}} (FX \Rightarrow X) \times FX$$
$$\xrightarrow{\text{(iv)}} X$$

where arrow (i) is derived from the diagonal on $FX \Rightarrow X$, (ii) from the strength, while (iii) and (iv) are just evaluations.

Towards the definition of $\mathtt{build}$, for a paranatural $\varphi : A \times (FX \Rightarrow X) \longrightarrow X$ take

$$\mathtt{build}(\varphi) \ : \ A \xrightarrow{A \times \ulcorner \mu \urcorner_!} A \times (FM_F \Rightarrow M_F)$$
$$\xrightarrow{\varphi M_F} M_F$$

Composing the above two definitions, one gets the commutative square

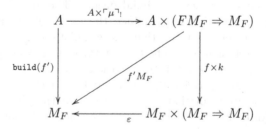

Since $k \cdot \ulcorner \mu \urcorner = \ulcorner \mathrm{id}_M \urcorner$, the path around the square reduces to f, and yields $\mathtt{build}(f') = f$.

The converse $\texttt{build}(\varphi)' = \varphi$ is the point-free version of lemma 1. It amounts to proving that the paranaturality of φ implies (indeed, it is equivalent) to the commutativity of

$$
\begin{array}{ccc}
A & \xrightarrow{\;A \times \ulcorner \mu \urcorner_!\;} & A \times (FM_F \Rightarrow M_F) \\
\Big\downarrow{\scriptstyle \widetilde{\varphi}X} & & \Big\downarrow{\scriptstyle \varphi M_F} \\
(FX \Rightarrow X) \Rightarrow X & \xleftarrow[\;\llparenthesis \kappa \rrparenthesis\;]{} & M_F
\end{array}
$$

where $\widetilde{\varphi}X$ is the transpose of φX. Showing this is an exercise in cartesian closed structure. On the other hand, the path around the square is easily seen to be $\texttt{build}(\varphi)'_X$.

To establish isomorphism (13), we internalize 15 similarly like we did 14 above. The natural correspondences

$$(-)' : \mathbb{C}(N_F, B) \longrightarrow \mathsf{Para}\,(X \times (X \Rightarrow FX),\ B)$$
$$\texttt{build} : \mathsf{Para}\,(X \times (X \Rightarrow FX),\ B) \longrightarrow \mathbb{C}(N_F, B)$$

are defined

$$
\begin{aligned}
g'_X\ :\ X \times (X \Rightarrow FX) &\xrightarrow{\;X \times \ell\;} X \times (X \Rightarrow N_F) \\
&\xrightarrow{\;\varepsilon\;} N_F \\
&\xrightarrow{\;g\;} B
\end{aligned}
$$

and

$$
\begin{aligned}
\texttt{build}(\psi)\ :\ N_F &\xrightarrow{\;N_F \times \ulcorner \nu \urcorner_!\;} N_F \times (N_F \Rightarrow FN_F) \\
&\xrightarrow{\;\psi N_F\;} B
\end{aligned}
$$

for $g : N_F \longrightarrow B$ and $\psi : X \times (X \Rightarrow FX) \longrightarrow B$. The arrow $\ell : (X \Rightarrow FX) \longrightarrow (X \Rightarrow FX)$ maps the coalgebra structures $x : X \to FX$ to the anamorphisms $\llbracket x \rrbracket : X \to N_F$. $\qquad\square$

References

1. Anlauff, M., Pavlovic, D., Waldinger, R., Westfold, S.: Proving authentication properties in the protocol derivation assistant. In: Degano, P., Küsters, R., Vigano, L. (eds.) Proceedings of FCS-ARSPA 2006. ACM (2006)
2. Bainbridge, E.S., Freyd, P.J., Scott, P.J., Scedrov, A.: Functorial polymorphism. Theor. Comput. Sci. **70**(1), 35–64 (1990). orrigendum in 71(3), 431

3. Bird, R., Meertens, L.: Nested datatypes. In: Jeuring, J. (ed.) MPC 1998. LNCS, vol. 1422, pp. 52–67. Springer, Heidelberg (1998). https://doi.org/10.1007/BFb0054285

4. Carboni, A., Freyd, P.J., Scedrov, A.: A categorical approach to realizability and polymorphic types. In: Main, M., Melton, A., Mislove, M., Schmidt, D. (eds.) MFPS 1987. LNCS, vol. 298, pp. 23–42. Springer, Heidelberg (1988). https://doi.org/10.1007/3-540-19020-1_2

5. Chadha, R., Kanovich, M.I., Scedrov, A.: Inductive methods and contract-signing protocols. In: Reiter, M.K., Samarati, P. (eds.) CCS 2001, Proceedings of the 8th ACM Conference on Computer and Communications Security, Philadelphia, Pennsylvania, USA, 6–8 November 2001, pp. 176–185. ACM (2001)

6. Freyd, P.: Algebraically complete categories. In: Carboni, A., Pedicchio, M.C., Rosolini, G. (eds.) Category Theory. LNM, vol. 1488, pp. 95–104. Springer, Heidelberg (1991). https://doi.org/10.1007/BFb0084215

7. Freyd, P.J.: Structural polymorphism. Theor. Comput. Sci. 115(1), 107–129 (1993)

8. Freyd, P.J., Girard, J.-Y., Scedrov, A., Scott, P.J.: Semantic parametricity in polymorphic lambda calculus. In: Proceedings Third Annual Symposium on Logic in Computer Science, pp. 274–279. IEEE Computer Society Press, July 1988

9. Gill, A., Launchbury, J., Peyton-Jones, S.: A short cut to deforestation. In: Proceedings of FPCA 1993. ACM (1993)

10. Girard, J.Y., Lafont, Y., Taylor, P.: Proofs and Types. Cambridge Tracts in Theoretical Computer Science. Cambridge University Press, Cambridge (1989)

11. Krstić, S., Launchbury, J., Pavlović, D.: Categories of processes enriched in final coalgebras. In: Honsell, F., Miculan, M. (eds.) FoSSaCS 2001. LNCS, vol. 2030, pp. 303–317. Springer, Heidelberg (2001). https://doi.org/10.1007/3-540-45315-6_20

12. Lincoln, P., Mitchell, J., Mitchell, M., Scedrovy, A.: Probabilistic polynomial-time equivalence and security analysis. In: Wing, J.M., Woodcock, J., Davies, J. (eds.) FM 1999. LNCS, vol. 1708, pp. 776–793. Springer, Heidelberg (1999). https://doi.org/10.1007/3-540-48119-2_43

13. Pavlovic, D.: Maps II: chasing diagrams in categorical proof theory. J. IGPL 4(2), 1–36 (1996)

14. Scedrov, A.: A guide to polymorphic types. In: Odifreddi, P. (ed.) Logic and Computer Science. LNM, vol. 1429, pp. 111–150. Springer, Heidelberg (1990). https://doi.org/10.1007/BFb0093926

15. Scedrov, A., Canetti, R., Guttman, J.D., Wagner, D.A., Waidner, M.: Relating cryptography and cryptographic protocols. In: 14th IEEE Computer Security Foundations Workshop (CSFW-14 2001), Cape Breton, Nova Scotia, Canada, 11–13 June 2001, pp. 111–114. IEEE Computer Society (2001)

16. Seldin, J.P., Hindley, J.R., Curry, T.H.B. (eds.): Essays on Combinatory Logic. Lambda Calculus and Formalism. Academic Press, London (1980)

17. Strachey, C.: Fundamental concepts in programming languages, lecture notes for the international summer school in computer programming. Copenhagen, August 1967

18. Wadler., P.: Theorems for free! In: Proceedings of FPCA 1989. ACM (1989)

There's No Time, The Problem of Conceptualising Time

Tajana Ban Kirigin[1]($\boxtimes$) and Benedikt Perak[2]

[1] Department of Mathematics, University of Rijeka, Rijeka, Croatia
bank@math.uniri.hr
[2] Faculty of Humanities and Social Sciences, University of Rijeka, Rijeka, Croatia
bperak@ffri.uniri.hr

Abstract. Among his numerous collaborations, over the last decade, Andre Scedrov has formed quite a stable research group. This would not have been possible without his leading scientific role, which has been equally measured by his generous personality, hospitality, and kindness. Among the obtained results and mathematical solutions, the group is particularly fond of real-time abstractions, called Circle-Configurations, which provide a way of handling both density as well as infinity of time in the model. Having in mind the broadness and variety of Andre Scedrov's interests, we hope to offer here yet another view on these constructions and formal timed models, enriched with the cognitive science perspective.

1 Real-Time Multiset Rewriting

Over a number of years our research group centered around Andre Scedrov has been developing multiset rewriting (MSR) models for various applications. Among the underlying mathematical solutions, we have been particularly fond of *circle-configurations*, the abstractions used in real-time MSR models [7,8].

These abstractions have been essential in providing some of the complexity results, *e.g.*, in protocol security [8,16]. However, the abstractions themselves may have been overshadowed by the concrete applications of the obtained results and may have not been fully appreciated for their mathematical elegance and meaning. At the same time, discussions with colleagues from different scientific fields, seem to provide some additional and unexpected appreciation for our mathematical constructions. In particular, there appears to be some relation to the conceptualization of time in the cognitive science and system theory. Some of the parallels in the approach to modelling time are presented here.

Modeling Time

Real time has explicitly been introduced to MSR models [7,8] by timestamping the atomic formulas, called *facts*, with real numbers, written $F@t$. A special predicate $Time$ is used to denote the global time. Multisets of facts containing

© Springer Nature Switzerland AG 2020
V. Nigam et al. (Eds.): Scedrov Festschrift, LNCS 12300, pp. 61–68, 2020.
https://doi.org/10.1007/978-3-030-62077-6_5

one occurrence of a *Time* fact represent system configurations, while MSR rules represent change in the system.

Time advancement is modelled by the *Tick* rule, $Time@T \longrightarrow Time@(T+\varepsilon)$, where ε can be instantiated by any positive real number formalizing the natural continuous aspect of time we experience in our everyday life. The remaining rules are instantaneous and denote possible events in the system.

Additional timing aspects are introduced through time constraints, which are comparisons of two time variables that may be attached to rules and system configurations, expressing conditions and system properties involving time. For example, $Time@T, Deadline(i, no)@T' \mid T' = T + 10$ specifies a critical configuration when the deadline for the unfinished task i will be reached in 10 time units. Reachability problems search for traces, *i.e.*, sequences of configurations, $S_I \longrightarrow^* S_G$, obtained by consecutive rule application. Traces in real-time models involve time variables and timestamps that range over real-time domain. The challenge here is to address density of time, the Zeno type phenomena, and the unboundedness of time, as there is no upper bound on the values of timestamps. All of these issues are handled using abstractions.

Abstracting Time

Circle-configurations are simple, finite sequences of symbols selected from a finite alphabet, containing no notion of time. By introducing these abstractions, traces over real-time MSR configurations are simulated with traces that do not contain any real numbers.

Abstractions are defined w.r.t. to a *resolution* parameter, D_{max}, extracted from the reachability problem at hand, and are formed by considering integer and decimal parts of timestamps of facts.

Definition 1 (Circle-Configurations). *Let $\mathcal{R}$ be a timed MSR with dense time, $\mathcal{GS}$ a goal, $\mathcal{CS}$ a critical configuration specification and S_0 an initial configuration. Let D_{max} be an upper bound on the numeric values appearing in $\mathcal{R}$, $\mathcal{GS}$, $\mathcal{CS}$ and S_0, and $S = F_1@t_1, F_2@t_2, \ldots, F_n@t_n, Time@t$.*

The pair $\mathcal{A}_S = \langle \Delta_S, \mathcal{U}_S \rangle$ is the circle-configuration of the configuration S defined as follows. The δ-configuration of S, Δ_S, is:

$$\Delta_S = \left\langle \{P_1^1, \ldots, P_{m_1}^1\}, \delta_{1,2}, \{P_1^2, \ldots, P_{m_2}^2\}, \delta_{2,3}, \ldots, \delta_{j-1,j}, \{P_1^j, \ldots, P_{m_j}^j\} \right\rangle$$

where $\{P_1^1, \ldots, P_{m_1}^1, P_1^2, \ldots, P_{m_j}^j\} = \{F_1, \ldots, F_n, Time\}$, timestamps of facts $P_1^i, \ldots, P_{m_i}^i$ have the same integer part, t^i, $\forall i = 1, \ldots, j$, and

$$\delta_{i,i+1} = \begin{cases} t^{i+1} - t^i, & \text{if } t^{i+1} - t^i \leq D_{max} \\ \infty, & \text{otherwise} \end{cases}, \quad i = 1, \ldots, j-1.$$

Unit circle of S is

$$\mathcal{U}_S = [\{Q_1^0, \ldots, Q_{m_0}^0\}z, \{Q_1^1, \ldots, Q_{m_1}^1\}, \ldots, \{Q_1^k, \ldots, Q_{m_k}^k\}],$$

where $\{Q_1^0, \ldots, Q_{m_0}^0, Q_1^1, \ldots, Q_{m_k}^k\} = \{F_1, \ldots, F_n, Time\}$, timestamps of facts in the same class, $Q_1^i, \ldots, Q_{m_i}^i$ have the same decimal part, $\forall i = 0, \ldots, k$, timestamps of facts $Q_1^0, \ldots, Q_{m_0}^0$ are integers, and the classes are ordered in the increasing order, i.e., $dec(Q_i^l) < dec(Q_j^{l'})$, for all $i \neq j$, where $1 \leq i \leq m_l$, $1 \leq j \leq m_{l'}$, $0 \leq l \leq k$, $1 \leq l' \leq k$.

Each circle-configuration is simply a pair of finite sequences of symbols, each sequence partitioning and ordering its facts. One sequence, δ-configuration, is formed according to the integer part, while the other, unit circle, relates to the decimal part of the timestamps. Unit circles can be clearly visualized using a circle, as shown in Fig. 1. For example, with $D_{max} = 3$, the circle-configuration of the configuration

$\{\ M@3.00000001,\ R@3.14,\ P@4.2,\ Time@111.2,\ Q@112.5333,\ S@114\ \}$

consists of the δ-configuration $\langle \{M,R\}, 1, \{P\}, \infty, \{Time\}, 1, \{Q\}, 2, \{S\} \rangle$ and the unit circle $[\{S\}_Z, \{M\}, \{R\}, \{P, Time\}, \{Q\}]$ as illustrated in Fig. 2.

$Q_1^j, \ldots, Q_{m_j}^j$ $Q_1^0, \ldots, Q_{m_0}^0$

$Q_1^1, \ldots, Q_{m_1}^1$

$Q_1^i, \ldots, Q_{m_i}^i$

$\langle \{M,R\}, 1, \{P\}, \infty, \{Time\}, 1, \{Q\}, 2, \{S\} \rangle$

Fig. 1. Unit circle   **Fig. 2.** Circle-configuration

Although these symbolic representations of configurations do not contain any real numbers, they provide enough information related to satisfaction of time constraints, which is necessary *e.g.*, for rule application.

Theorem 1 (Bisimulation). *Let T be a reachability problem for a balanced system with facts of bounded size. Let D_{max} be an upper bound on the numeric values in T. Then $S_I \longrightarrow^* S_G$ for some configurations S_I and S_G in T if and only if $\mathcal{A}_I \longrightarrow^* \mathcal{A}_G$, where $\mathcal{A}_I$ and $\mathcal{A}_G$ are, respectively, the circle-configurations of S_I and S_G w.r.t. D_{max}.*

The systems to which the above theorem relates, involve configurations with a fixed number of facts, each containing a bounded number of symbols. Such systems typically represent systems with a fixed, *i.e.*, bounded memory. Specifically, balanced systems only have rules with the same number of facts on each side of rewrite rules. Relaxing these conditions leads to undecidability of the reachability problem for MRS models [3,9]. For balanced real-time MSR models with a bound on the size of facts, the above bisimulation result allows the search for solutions of reachability problems symbolically, not necessarily involving explicit values of the timestamps, i.e., the real numbers. Circle-configurations provide a sound and complete representation of traces:

$$\mathcal{S}_I \to_{r_1} \cdots \to_{r_{i-1}} \mathcal{S}_{i-1} \to_{r_i} \mathcal{S}_i \to_{r_{i+1}} \cdots \to_{r_n} \mathcal{S}_G$$
$$\updownarrow \qquad\qquad\qquad \updownarrow \qquad \updownarrow \qquad\qquad\qquad \updownarrow$$
$$\mathcal{A}_I \to_{(r'_1)} \cdots \to_{(r'_{i-1})} \mathcal{A}_{i-1} \to_{(r'_i)} \mathcal{A}_i \to_{(r'_{i+1})} \cdots \to_{(r'_n)} \mathcal{A}_G$$

In the above bisimulation, (r'_i) is a representation of MSR rules over abstractions. In particular, to an instantaneous MSR rule corresponds a single rule over abstractions.

A *Tick* rule (for any ε!) is simulated with a finite sequence of *Next* rules over circle-configurations. There are eight cases defining the *Next* rule, exactly one of which is applicable to a given circle-configuration. The fact *Time* is moved clockwise on the unit-circle, updating the δ-configuration when needed, as illustrated below:

$\Delta = \langle\ldots, \mathcal{P}_{-1}, \delta_{-1}, \{Time, Q_1, \ldots, Q_m\}, \delta_1, \mathcal{P}_1, \ldots, \mathcal{P}_k\rangle$ $\Delta' = \langle\ldots, \mathcal{P}_{-1}, \delta_{-1}, \{Q_1, \ldots, Q_m\}, 1, \{Time\}, \delta_1 - 1, \mathcal{P}_1, \ldots, \mathcal{P}_k\rangle$

Sequences of circle-configurations obtained by consecutive applications of *Next* rules simulate the "flow of time". They represent a series of "atomic" changes in the system where the fact *Time* moves just enough so that a different circle-configuration is obtained. This can be observed as a change in the system, substantial enough to reflect on the system behaviour and properties which are specified through the satisfiabiliy of time constraints. For example, the circle-configuration above on the right represents a configuration in which $Time@T_1$ has the same decimal part as $F_1@T_2$, satisfying, hence, some time constraint of the form $T_1 = T_2 \pm D$. This constraint is not satisfied by the previous circle-configuration, which instead satisfies some time constraint $T_1 < T_2 \pm D$. On the other hand, it may be the case that $Time@T \longrightarrow Time@(T + \varepsilon)$ is applied for small ε so that both the enabling and the resulting configurations correspond to the same circle-configuration. This means, that w.r.t. time constraints, *i.e.*, time conditions, the time increase is not noticeable in the system.

Abstractions, therefore, represent a way of removing the dimension of time from a system, so that time ticks are replaced by (possibly empty) sequences of distinct abstractions, denoting noticable change in the system. For more details on real-time MSR see [8].

2   Conceptualization of Sequence as Time

The above abstractions imply the sequential conceptualization of time that contests the usual folk-based notion of the time as an objective ontological category. What we assume by folk theory is the common metaphorical conception of time as resource ("We're out of time"), as container ("We did it in five days"), as person ("Time is a great teacher"), as an object in motion ("Time is passing quickly") or as a static reference point where the conceptualizer is moving towards ("We are moving towards future"), or with earlier events in front of later events ("Leave the history behind you!") [2,11].

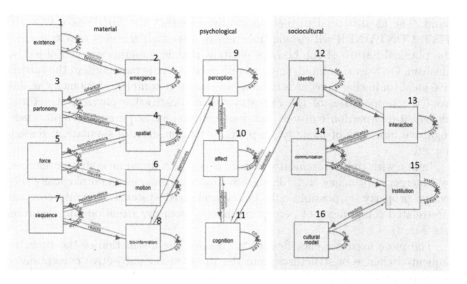

Fig. 3. Hierarchical structure of complex dynamic system organization. [12]

The sequential nature of the system changes is the basis for the folk conception of the ontologically objective existence of time: time exists. The reoccurring system changes are also the source domain for figurative extension of the time units: day, month, year. This apparent correlation of time and sequences of physical changes yielded two opposing philosophical views with the claim of the ontologically objective nature of the time.

The first one is the relational perspective that considers time as dependent on the sequences of the physical events in the universe. According to this position, promoted by Leibniz, Mach and Saint Augustine, there is no existence of the time outside the succession of events. On the other hand, the opposing philosophical realist view, promoted by Isaac Barrow and Isaac Newton, claims that time exists independently of physical occurrences and that it would exist even in an empty universe. The relational perspective conceptualizes the TIME as having a SEQUENTIAL CHANGE, while the realists metaphorically conceptualize TIME as a CONTAINER FOR SEQUENTIAL CHANGE. Both conceptualizations pragmatically imply existence of the TIME entity irrespective of the cognizer that can perceive the sequential change.

In contrast to this metaphorical conceptualization of time as an ontological objective entity, systemic approach assumes that a time emerges as an psychological phenomenon referring to the process of subjective perception, experience and conceptualization of sequences. This position is similar to the idealist view, partly described by Saint Augustine, Immanuel Kant and many other physicist and philosophers. According to the idealist perspective the time is a construction of the human mind, and correspondingly the existence of time depends on the existence of the human observer. Where the systemic view potentially differs from the idealist perspective is in the conceptualization of the observing mind. The idealist tend to metaphorically construct the MIND as a COGNI-TIVE CONTAINER with some (inherent or essential) features that transcend the physical nature of the body, a position that is sometimes called objective idealism. On the other hand, the systemic perspective is more akin to the subjective idealism in that it regards mind as an emergent complex phenomena arising from the propensities of the organization of constitutive physical structures, biological-information features, sensors, evolutionary acquired affect processes, cognitive functions of memory, mapping, attention, mental simulation, reasoning, etc.

In line with the hierarchical structure of complex dynamic systems organized by Emergence principle [1,4,13], the emergence of time, as an ontologically subjective property, is possible only in self-organized systems of highly complex interrelated components of perception, affect, memory, attention and cognition (see Fig. 3).

The more accurate classification argues for the distinction of the $time_1$ as sequential change of structures from the $time_2$ as the subjective perception of the sequential change.

$Time_1$ can be defined as the sequential change of some system (Fig. 3, level 7+). It is a dynamic process of a material entity, endowed at least with the features of the existence, emergent spatial structuring of partonymy and dynamic forces that:

- undergoes a change in its internal organization as a whole or
- is a part of changing features with regards to some external material structure while maintaining its structural complexity.

$Time_1$ has ontologically and epistemologically objective features, meaning that sequences of change within a system are (mostly) observer independent. The sequences occur regardless of the perception by any cognizer.

$Time_2$, on the contrary, involves cognition of the sequential change by some cognizer (Fig. 3, level 11+). The cognizer is a system that maintains the internal structure while undergoing sequential changes. It is endowed with at least the ontologically objective features of:

- maintaining the complexity features while interacting with the environment,
- internal storing of the data created by the interaction, and
- ontologically subjective emergent features of perception, affection processes that signal the quality of the perceived structures, and cognition processes that allow for the mental simulation and evaluation of the stored information about the quality of the perceptions.

The cognizer observes the sequential changes, as a perception:

- of the internal change within cognizers system (body changes, interception) and/or
- of the external structure changes outside cognizers system (visual, auditive perception, exteroception).

Time$_2$ has, obviously, a subjective character. In fact, time$_2$ is the subjective perception of time$_1$. Due to the varying levels of dynamic stability within the system, the emergent properties of subjective time are prominently correlated with the cognitive functions and traits. Different species will differ in perception of subjective time$_2$ due to the internal representation system, different sensors, attention and memory processing systems [10].

Even among individuals of the same species, for instance humans, processing perception of sequential change, time$_1$, is dependent of the state of the cognition faculties. The obvious example is the state of deep sleep, without dreams, non-REM stage, where there is no cognition of external perception, and no cognition of internal mental simulation, and subsequently no perception of time. So, if there is no perception of time is there time at all?[1]

At least, if there is a (perception of) sequence, there is time.[2]

How is it that we all use time concepts in interpersonal setting and even have global objective time structures? As for the global time, it is clear that the time measured by some clock, be it atomic, quartz or mechanical is just a standardized physical measure of some sequential process, much like one kilo is a standardized unit for measuring mass. Regarding the similar experience of subjective time, it seems that the evolutional predispositions of the neural structures facilitate the approximation of the subjective time in a commensurable way.

This brings us to the question of the scientific terminology that uses the notion of time. Are we talking about the sequential change or perception of sequences? What is really the concept of time that is referred to in Einstein's relativity theory? Gödel argued that there cannot be an objective lapse of time [5,15]. When theoretical physicist Carlo Rovelli [6,14], claims that "time is an illusion" he is rightly observing that naive perception of time flow doesn't correspond to the physical reality, but can we say the time is non-existent?

3   Relating the Approaches to Time

When we escape into the formal mathematical world we can claim there's no time in traces over our circle-configurations. There are just sequences of symbolic abstractions that contain no explicit real-time values. Still, such traces of subsequent applications of the *Next* rule simulate the usual, naive flow of time in a sound and complete way, providing an (un-timed, but time-like) representation of real-time.

However, by carefully analyzing the solution that uses abstractions, one can suspect that there are some elements that represent subjective perception. Namely, the definition of circle-configurations relates to a bounded (balanced) system and the observation of such system (extracting the parameter D_{max}). These elements may involve the perception of time within the system. The flow of time in the given system is therefore specific, *i.e.*, subjective, and is reduced to a sequence of such system-specific abstractions.

---

[1] Is it time for change?.

[2] The perception of time is the perception of change.

Acknowledgments. This work has been supported in part by the Croatian Science Foundation under the project UIP-05-2017-9219 and the University of Rijeka under the project Initial Grants 1016-2017.

References

1. Bar-Yam, Y.: Dynamics of Complex Systems. CRC Press, Boca Raton (2019)
2. Brdar, M., Brdar-Szabó, R., Perak, B.: Metaphor repositories and cross-linguistic comparison. In: Metaphor and Metonymy in the Digital Age: Theory and Methods for Building Repositories of Figurative Language, vol. 8, p. 64 (2019)
3. Durgin, N.A., Lincoln, P., Mitchell, J.C., Scedrov, A.: Multiset rewriting and the complexity of bounded security protocols. J. Comput. Secur. **12**(2), 247–311 (2004)
4. Emmeche, C., Køppe, S., Stjernfelt, F.: Explaining emergence: towards an ontology of levels. J. Gen. Philos. Sci. **28**(1), 83–117 (1997)
5. Gödel, K.: An example of a new type of cosmological solutions of Einstein's field equations of gravitation. Rev. Mod. Phys. **21**(3), 447 (1949)
6. Jaffe, A.: The illusion of time. Nature **556**(7701), 304–306 (2018)
7. Kanovich, M., Kirigin, T.B., Nigam, V., Scedrov, A., Talcott, C.: Discrete vs. dense times in the analysis of cyber-physical security protocols. In: Focardi, R., Myers, A. (eds.) POST 2015. LNCS, vol. 9036, pp. 259–279. Springer, Heidelberg (2015). https://doi.org/10.1007/978-3-662-46666-7_14
8. Kanovich, M.I., Ban Kirigin, T., Nigam, V., Scedrov, A., Talcott, C.L.: Time, computational complexity, and probability in the analysis of distance-bounding protocols. J. Comput. Secur. **25**(6), 585–630 (2017)
9. Kanovich, M.I., Ban Kirigin, T., Nigam, V., Scedrov, A., Talcott, C.L., Perovic, R.: A rewriting framework and logic for activities subject to regulations. Math. Struct. Comput. Sci. **27**(3), 332–375 (2017)
10. Matthews, W.J., Meck, W.H.: Temporal cognition: connecting subjective time to perception, attention, and memory. Psychol. Bull. **142**(8), 865 (2016)
11. Núñez, R.E., Sweetser, E.: With the future behind them: convergent evidence from aymara language and gesture in the crosslinguistic comparison of spatial construals of time. Cogn. Sci. **30**(3), 401–450 (2006)
12. Perak, B.: The role of the metonymy and metaphor in the conceptualization of nation. an emergent ontological analysis of syntactic-semantic constructions. In: Metaphors in the Discourse of the National. John Benjamins (2019)
13. Perak, B., D'Alessio, S.P.: Culture as an emergent property of the embodied cognition. In: Avanture kulture: kulturalni studiji u lokalnom kontekstu. Jesenski i Turk (2013)
14. Rovelli, C.: The Order of Time. Riverhead Books (2019)
15. Schilpp, P.A., Gödel, K.: A remark about the relationship between relativity theory and idealistic philosophy (1949)
16. Urquiza, A.A., et al.: Resource-bounded intruders in denial of service attacks. In: 2019 IEEE 32nd Computer Security Foundations Symposium (CSF), pp. 382–38214. IEEE (2019)

Andre and the Early Days of Penn's Logic and Computation Group

Dale Miller[✉]

Inria & LIX, École Polytechnique, Palaiseau, France
Dale.Miller@inria.fr

I first met Andre Scedrov in the Fall of 1983 when I joined the Computer and Information Science Department faculty at the University of Pennsylvania. Andre had started in the Mathematics Department at Penn the year earlier. As was apparent even then, Andre's approach to doing research lead him to seek out colleagues. In 1983, he crossed the divide between the Mathematics Department and the Computer and Information Science Departments—that is, he crossed 33rd Street in West Philadelphia—in search of joint research projects with computer scientists.

At that time, Andre and Peter Freyd were holding a weekly *Geometric Logic Seminar* in the Mathematics Department. Andre and Peter invited me to speak there twice in 1984. Around that time, it was decided to expand the seminar's scope and to rename it as the *Penn Logic Seminar*. In a proposal that Andre and I submitted to the NSF in December 1986, we described the three-year-old seminar by saying that it was "held jointly between the Mathematics and Computer Science Departments. There is also substantial involvement from the Philosophy and Linguistics Departments at Penn as well as from various departments of local universities. The attendees for this seminar include the following Penn faculty: Peter Freyd and Scedrov of the Mathematics Department, Peter Buneman, Jean Gallier, Saul Gorn, Aravind Joshi and Miller of the Computer and Information Science Department, Scott Weinstein and Zoltan Domotor of the Philosophy Department, and Henry Hiz of the Linguistics Department." Eventually, we renamed the seminar once more to be the Logic and Computation Seminar.

In 1986, Albert Meyer organized the first IEEE Symposium on Logic in Computer Science (LICS) at MIT in Cambridge. The establishment of that conference series played a major role in shaping how many of us at Penn understood the core of our research goals. LICS 1986 brought together a large number of famous logicians, mathematicians, and computer scientists, and it helped us see ourselves as being involved in a new and vital topic. It is notable, however, that no one from Penn was involved in the organization of LICS 1986, nor did we have any accepted papers there.

The first couple of years of this seminar were used to educate ourselves on several topics, including polymorphic λ-calculus, category theory, linear logic, logic programming, denotational semantics, and automated deduction. As I remember it, Andre was instrumental in organizing many the faculty, postdocs, graduate students, and visitors into a steady stream of lectures on foundational topics.

© Springer Nature Switzerland AG 2020
V. Nigam et al. (Eds.): Scedrov Festschrift, LNCS 12300, pp. 69–70, 2020.
https://doi.org/10.1007/978-3-030-62077-6_6

The seminar was usually held on the second floor of the David Rittenhouse Laboratory, which hosted the Mathematics Department. Attendance was typically strong: we all had the sense that there was a great deal of dynamism in this topic. The appearance in the late 1980s of both linear logic and the π-calculus helped to convince us that this topic was open to fresh and important shifts in perspectives. No longer were we only attempting to apply well known and mature logic techniques to computing, computing itself was influencing foundations, even the foundations of logic.

This education phase for the Logic and Computation Group was very successful. For example, people from Penn started to have papers appearing in LICS. In the second LICS meeting in 1987, three accepted papers were authored by attendees of this seminar, and Andre co-authored two of those papers. In fact, during the five years 1987–1991, Andre authored six papers accepted at LICS for which he had 12 co-authors. Andre also served as the PC Chair for LICS 1992.

In 1987, the Logic and Computation Group grew with the hiring of Carl Gunter and Val Tannen by the Computer and Information Science Department. Around that time, we had posters printed and mailed that advertised our interdisciplinary approach to Logic and Computation. We felt at that time that our main competitor was the Laboratory for Foundations of Computer Science (LFCS) at the University of Edinburgh. A year or so after we distributed our poster, Carnegie Mellon University formed a similar interdisciplinary group and distributed their poster, an event that we referred to as "The Empire Strikes Back."

Andre was always well connected to many other researchers. For example, the group had a series of visitors who were collaborators of Andre: Harvey Friedman, Jean-Yves Girard, Andreas Blass, Max Kanovich, Phil Scott, Mitsuhiro Okada, Jim Lipton, etc. In addition, Andre's Erdös number is 2 and, as a result, many people at Penn have an Erdös number of 3.

I owe a big thanks to Andre. It was immensely valuable for me in 1983, as a beginning researcher in an interdisciplinary setting, to meet and collaborate with Andre. He was warm and willing to helped me, my students, and our colleagues. His assistance also extended to providing sage advice and guidance with maneuvering the professional and academic world of universities and funding agencies.

Thank you, Andre, for having been there at the beginning of my professional life. You have helped me find and develop an enthusiasm for the interdisciplinary research that has been with me my full career.

Formal Verification of Ethereum Smart Contracts Using Isabelle/HOL

Maria Ribeiro[1], Pedro Adão[1,2(✉)], and Paulo Mateus[1,2]

[1] Instituto Superior Técnico, Universidade de Lisboa, Lisbon, Portugal
{maria.ribeiro,pedro.adao}@tecnico.ulisboa.pt,
pmat@math.tecnico.ulisboa.pt
[2] Instituto de Telecomunicações, Lisbon, Portugal

Abstract. The concept of blockchain was developed with the purpose of decentralizing the trade of assets, suppressing the need for intermediaries during this process, as well as achieving a digital trust between parties. A blockchain consists in a public immutable ledger, constituted by chronologically ordered blocks such that each block contains records of a finite number of transactions.

The Ethereum platform, that this paper builds upon, is implemented using a blockchain architecture and introduces the possibility of storing Turing complete programs. These programs, also known as smart contracts, can then be executed using the Ethereum Virtual Machine. Despite its core language being the EVM bytecode, they can also be implemented using a higher-level language that is later compiled to EVM, being Solidity the most used. Among its applications stand out decentralized information storage, tokenization of assets, and digital identity verification.

In this paper we propose a method for formal verification of Solidity smart contracts in Isabelle/HOL. We start from the imperative language and big-step semantics proposed by Schirmer [23], and adapt it to describe a rich subset of Solidity, implementing it using the Isabelle/HOL proof assistant. Then, we describe the properties about programs using Hoare logic, and present a proof system for the language, for which results on soundness and (relative) completeness are obtained.

Finally, we describe the verification of an electronic voting smart contract, which illustrates the degree of proof complexity that can be achieved using this method. Examples of smart contracts containing overflow and reentrancy vulnerabilities are also presented.

Keywords: Formal verification · Isabelle/HOL · Hoare logic · Smart contracts · Solidity · Ethereum

---

Partially supported by Programa Operacional Competitividade e Internacionalização (COMPETE 2020), Fundo Europeu de Desenvolvimento Regional (FEDER) through Programa Operacional Regional de Lisboa (Lisboa 2020), Project BLOCH - LISBOA-01-0247-FEDER-033823, and Fundação para a Ciência e Tecnologia (FCT) project UID/EEA/50008/2019.

V. Nigam et al. (Eds.): Scedrov Festschrift, LNCS 12300, pp. 71–97, 2020.
https://doi.org/10.1007/978-3-030-62077-6_7

1   Introduction

The emergence of the blockchain concept was associated with the appearance of Bitcoin, one of the first decentralized cryptocurrencies, introduced in 2008 by Satoshi Nakamoto [21]. A cryptocurrency is independent of any central administrative entities and uses instead a *peer-to-peer* digital system, managed by a network of nodes. Transactions are stored in a blockchain, an append-only public ledger, through the process of *mining*. Nodes in the network, also known as *miners*, try to solve a difficult computational problem called *proof-of-work*. When a transaction is verified by the network it is incorporated into the blockchain using a cryptographic hash function, which includes data from the previous block's hash and makes the whole chain cryptographically secure and, therefore, immutable.

Our work focuses on the Ethereum platform, proposed by Vitalik Buterin [4] in 2013, which similarly uses a blockchain architecture but also introduces the feature of storing Turing complete programs, known as *smart contracts*. These programs can be executed by the stack-based Ethereum Virtual Machine (EVM), and its formalization was first approached by Gavin Wood [26]. Ethereum also introduces the concept of *gas*, as each operation in the virtual machine has an associated cost in *ether*, the Ethereum currency. When a contract is executed, either by being called by a transaction or by code in another contract, the original transaction initiator needs to pay for the total cost of operations.

Given the valuable assets in these contracts, and the fact that they are immutable, studying the security of these programs becomes of uttermost importance. With that in mind, the main goal of this work is to introduce a formal verification method of Ethereum smart contracts using Isabelle, a higher order logic (HOL) theorem prover. We have chosen to verify smart contracts written in Solidity, a higher level language that compiles to EVM bytecode.

The main reference for our language, and respective semantics and proof system, is the work by Schirmer [23]. We adapt the proposed language for sequential programs to capture a relevant subset of Solidity. Our main additions were the modeling of Solidity calls, both internal and external, Solidity exceptions, and reverting all state modifications. To formalize the meaning of these new operations in terms of execution, the big-step semantics was extended. The verification of programs is done using Hoare logic. Soundness and (relative) completeness results for the proof system are presented.

The concept of weakest precondition [6] is presented and used both for optimizing program verification and for the completeness result. Regarding the first, and following the work by Frade and Pinto [7], we enhance the weakest precondition and verification condition computations with the cases for *Dyncom*, *Require* and *Init*. The proof of (relative) completeness, based on the proof by Winskel [25], is extended with the *Call, Handle, Revert, Dyncom, Require* and *Init* cases.

To conclude the paper we present some relevant examples of applications such as electronic voting, tokens, and reentrancy, describing and analyzing this way the expressiveness of the language.

Related Work. Previous efforts have been made by the research community to formally verify smart contracts. Hirai formalized the EVM semantics in Lem and used Isabelle/HOL to prove safety properties of Ethereum smart contracts [11]. Amani et al. [1] formalized the EVM semantics in Isabelle/HOL and proposed a sound program logic to verify correctness of smart contracts. Grishchenko et al. [9] formalized a complete small-step semantics of EVM bytecode in F*, and defined security properties for smart contracts such as call integrity and atomicity. Hildebrandt et al. [10] formalized the EVM semantics in the $\mathbb{K}$ framework. Bhargavan et al. [3] introduced a framework that translates smart contracts from Solidity to F*, allowing verification of functional correctness and safety, as well as a decompilation from EVM bytecode to F* for analysis of low-level properties.

As for analysis of Solidity code, Bartoletti et al. [2] proposed a calculus for a fragment of Solidity with a single primitive to transfer currency and invoke contract procedures, and Jiao et al. [14] developed a formal semantics for Solidity in the $\mathbb{K}$ framework that allows formal reasoning about high-level contracts. Zakrzewski [27] proposed a semantics for a small fragment of Solidity in Coq.

Some automatic analysis tools for analysing Ethereum smart contracts have also been developed as are the cases of Oyente [17], Maian [22], Mythril [20], and Securify [24]. A survey on these techniques and tools can be found in [8].

Andre's Influence in This Work. Scedrov's results on linear logic [15,16] significantly shaped our work. Linear logic encompasses the dynamics of algorithms and resources, and its main impacts have been in computer science rather than traditional mathematics. Linear logic significantly influenced the design of Hoare triples, which are the basis of this work. Moreover, the use of formal methods in Scedrov's work [18,19], namely on process algebras, has also been a significant contribution to the security area in general and inspired this work in particular. Indeed, this paper's primary goal is to present a proof-based method to derive security properties in an imperative language for contracts over a blockchain, which is a very restrictive form of concurrent programming, and for which we do not impose polynomial-time bounding. More importantly, Andre directly impacted the work and scientific career of two of the authors. Pedro Adão and Paulo Mateus were respectively PhD and Postdoc students of Andre.

2   The Ethereum Blockchain

Ethereum can be seen as a decentralized computing platform since it uses a blockchain architecture and introduces the feature of storing *smart contracts*.

In this section we present a simplified definition of the Ethereum blockchain. In the following definitions let $\mathbb{N}_x$ the set of non-negative integers with size up to x bits and $\mathbb{B}$ the set of bytes. An account is an object of the Ethereum environment that is identified by a 160-bit string known as the account's address.

Definition 1 (World state). *The world state is a mapping σ between addresses (160 bit strings) and account states.*

$$\sigma : \{0,1\}^{160} \;\rightarrow\; \mathbb{N}_{256} \times \mathbb{N}_{256} \times (\{0,1\}^{256} \rightarrow \{0,1\}^{256}) \times \mathbb{B}^*$$

There are two types of accounts: *externally owned accounts (EOA)* and accounts associated with code (*contract accounts*).

Definition 2 (Account state). *Given an address a, the account state $\sigma(a)$ is a tuple $\mathcal{A} = \langle nonce, balance, storage, code \rangle$, where*

- *$nonce \in \mathbb{N}_{256}$ is the nonce of the account. If a is the address of an EOA, corresponds to the number of transactions sent from this address. If a is the address of a contract account, corresponds to the number of contract-creations made by this account;*
- *$balance \in \mathbb{N}_{256}$ is the value of ether owned by account a;*
- *$storage$ is a mapping between 256-bit values and corresponds to the account's storage;*
- *$code \in \mathbb{B}^*$ is the EVM code of this account. In case of an EOA corresponds to the empty string.*

There are two types of transactions: *contract creation transactions* and *transactions which result in message calls*. A transaction is triggered by an external actor.

Definition 3 (Transaction). *A transaction is a tuple $T = \langle nonce, gasprice, gaslimit, from, to, value, init/data \rangle$, where*

- *$nonce \in \mathbb{N}_{256}$ is the number of transactions sent by address from;*
- *$gasprice \in \mathbb{N}_{256}$ is equal to the cost per unit of gas, in ether, for all computation costs of this transaction;*
- *$gaslimit \in \mathbb{N}_{256}$ is equal to the maximum amount of gas that should be used in the execution of this transaction;*
- *$from \in \{0,1\}^{160}$ is the address of the transaction's sender;*
- *$to \in \{0,1\}^{160}$ is the address of the transaction's recipient;*
- *$value \in \mathbb{N}_{256}$ is the value of ether to be transferred to the message call's recipient or, in the case of contract creation, as an endowment to the newly created account.*

Additionally, in the case of a contract creation transaction

- *$init$ is the EVM code for the account initialization procedure;*

In the case of a message call

- *$data$ is the input data of the message call.*

A *message call* is an internal concept which consists of data (a set of bytes) and value (specified as ether) passed from one account to another. It may be triggered by a transaction, where the sender is an EOA, or by EVM code, where the sender is a contract account.

Transactions are grouped and stored in finite blocks.

Definition 4 (Block). *A block B is a package of data constituted by*

- *a header, constituted by the block's number, timestamp, nonce, difficulty, beneficiary, state and hash of its parent's block header;*
- *a list of transactions* $\mathcal{T} = \{T_1, \ldots, T_m\}$.

The block's difficulty influences the time that it takes to find a valid nonce for the block and thus solving the proof-of-work mining problem. The beneficiary is the address which receives all the fees from the successful mining of this block. The fact that the hash of this block's header includes its parent's hash, is essential to the blockchain's immutability. The stored state corresponds to the one after all transactions are executed.

Ethereum can be seen as a transaction-based state machine. In such a representation a transaction represents a valid transition between two states σ_t and σ_{t+1}. Since transactions are grouped in finite blocks, a block may also represent a state transition σ_t' and σ_{t+1}'. These transitions between blocks introduce the concept of a chain of blocks, a blockchain.

Definition 5 (Blockchain). *A blockchain is defined as an ordered sequence of blocks* $\mathcal{B} = \{B_0, B_1, \ldots\}$.

In this paper we present an approach to the formal verification of Solidity smart contracts. Regarding code structure, a Solidity contract consists, as it follows a object-oriented structure, of a set of *state variables* which are part of the account's storage, and a set of *function declarations*. Functions in a contract can introduce local variables, which are stored in the memory. Solidity also contains a set of globally available variables that can be accessed regarding the current block, transaction, message call and address.

A function can be called by an external user, an EOA, which initiates a transaction, or by another contract. This happens when a called contract contains code that calls another contract, generating a new message call. A function call can be internal or external and an external call can be a *regular call* or a *delegate call*, in which case the code is executed in the context of the calling contract. Details about these methods and respective implementation are presented in Sect. 3.4.

Every contract has a *fallback function*, which is automatically executed whenever a call is made to the contract and none of its other functions match the given function identifier, or in the cases where no data is supplied.

Solidity also allows the usage of *exceptions*. Whenever an exception is thrown, all state changes are reverted. Our approach for modeling exceptions and *state reversion* is described in Sect. 3.4.

3 The SOLI Language

In this section we define the core elements of the language and introduce a set of big-step execution rules to describe their semantics. The main reference for our language and respective semantics is the work by Schirmer [23] which we adapt and extend to capture a relevant fragment of Solidity.

3.1 Syntax

The syntax of our language is a combination of deep and shallow embeddings. Commands are represented by an inductive, state dependent, datatype whereas some other syntactic elements are defined as abbreviations of their semantics. Boolean expressions, *bexp*, and assertions, *assn*, are defined as state sets.

Definition 6 (Syntax). *Let $'s$ be the state space type. The syntax for* boolean expressions and assertions *is defined by the types $'s$ bexp and $'s$ assn, respectively. The syntax for* commands *is defined by the polymorphic datatype $'s$ com, where $'s \Rightarrow 's$ is a state-update function and $fname$ the type of function names.*

$$
\begin{array}{lll}
's \ bexp & := & 's \ set \\
's \ assn & := & 's \ set \\
's \ com & := & \textbf{Skip} \mid \textbf{Upd} \ 's \Rightarrow 's \mid \textbf{Seq} \ 's \ com \ 's \ com \mid \\
& & \textbf{If} \ 's \ bexp \ 's \ com \ 's \ com \mid \textbf{While} \ 's \ bexp \ 's \ com \mid \\
& & \textbf{Dyncom} \ 's \Rightarrow 's \ com \mid \textbf{Call} \ fname \mid \textbf{Revert} \mid \\
& & \textbf{Handle} \ 's \ com \ 's \ com \mid \textbf{Require} \ 's \ bexp \mid \\
& & \textbf{Init} \ 's \ com \ 's \Rightarrow 's \Rightarrow 's
\end{array}
$$

Regarding the definition of commands, **Upd** is used to model assignments by executing a state-update function $'s \Rightarrow' s$. Conditional statements and while loops are defined with the usual syntax. The **Skip** command, which does nothing, is also defined.

In order to allow complex operations such as calling other functions and reverting all state changes, the following commands are introduced in SOLI. **Dyncom** is a command which receives a state and allows to write statements which are state dependent. This is useful when referring to states in different steps of execution. A general **Call** is introduced, which receives a function name. It corresponds to the simplest form of calling a procedure. The different types of procedure calls and respective execution details are described in detail in Sect. 3.4. **Revert** throws a revert type exception which signals that the state must be reverted, and **Handle** is an auxiliary command to handle state reversion if signaled. **Require** models Solidity exceptions, and **Init** models state reversion whenever a REVERT exception is thrown. Both commands are detailed in Sect. 3.4.

3.2 Concrete Syntax

To improve the readability of SOLI programs, some syntax translations are introduced. $\{|b|\}$ is defined as the set of states for which the predicate b holds. Syntax translations are defined as follows, where c_1 and c_2 are commands, b a boolean and s a state.

$$'x ::= a \rightharpoonup \mathbf{Upd} \ (\lambda s. \ s\langle\!\langle x := a \rangle\!\rangle)$$
$$c_1;\,;c_2 \rightharpoonup \mathbf{Seq} \ c_1 \ c_2$$
$$\mathbf{IF} \ b \ \mathbf{THEN} \ c_1 \ \mathbf{ELSE} \ c_2 \rightharpoonup \mathbf{If} \ \{\!|b|\!\} \ c_1 \ c_2$$
$$\mathbf{IF} \ b \ \mathbf{THEN} \ c_1 \rightharpoonup \mathbf{IF} \ b \ \mathbf{THEN} \ c_1 \ \mathbf{ELSE} \ \mathbf{Skip}$$
$$\mathbf{WHILE} \ b \ \mathbf{DO} \ c \rightharpoonup \mathbf{While} \ \{\!|b|\!\} \ c$$
$$\mathbf{REQUIRE} \ b \rightharpoonup \mathbf{Require} \ \{\!|b|\!\}$$

3.3 Semantics

To model big-step semantics, the state space $'s$ is augmented, as described by the datatype $'s \ state$, with information about whether exceptions were thrown.

$$'s \ state \ := \ \text{Normal} \ 's \ | \ \text{Rev} \ 's$$

To formalize the execution relation, the partial function Γ is introduced, which maps function names to the corresponding bodies.

In Isabelle such a function is defined as $'b \Rightarrow \ 'a \ option$ where $'a \ option = Some \ 'a \ | \ None$. In the case of Γ being defined for m, it is selected using the $(Some \ m) = m$.

Definition 7 (Big-step semantics). *The* big-step semantics *for SOLI is based on a deterministic evaluation relation formalized by the predicate*

$$\Gamma \vdash \langle c, \ s \rangle \Rightarrow t$$

where

$$\Gamma \ :: \ fname \rightharpoonup \ 's \ com$$
$$c \ :: \ 's \ com$$
$$s, \ t \ :: \ 's \ state$$

and evaluated accordingly to the set of rules represented in Fig. 1. The meaning for this predicate is as expected, that is, if command c is executed in initial state s, then the execution terminates in state t.

3.4 Additional Language Features

Gas isn't modeled in SOLI, the main reason is because it expresses a high level language and would not be accurate to measure gas consumption since it is defined for each opcode. Also the goal is to verify properties which are expressed in a symbolic way, and in most of the cases this measure is not relevant. One could however estimate bounds for SOLI commands, with the help of some side tools such as the Remix compiler, and defining the consumption inductively.

$$\frac{}{\Gamma \vdash \langle \textbf{Skip},\ \text{Normal } s \rangle \Rightarrow \text{Normal } s}\ (Skip) \qquad \frac{}{\Gamma \vdash \langle \textbf{Upd } f,\ \text{Normal } s \rangle \Rightarrow \text{Normal } (f\ s)}\ (Upd)$$

$$\frac{\Gamma \vdash \langle c_1,\ \text{Normal } s_1 \rangle \Rightarrow s_2 \qquad \Gamma \vdash \langle c_2,\ s_2 \rangle \Rightarrow s_3}{\Gamma \vdash \langle \textbf{Seq } c_1\ c_2,\ \text{Normal } s_1 \rangle \Rightarrow s_3}\ (Seq)$$

$$\frac{s \in b \qquad \Gamma \vdash \langle c1, \text{Normal } s \rangle \Rightarrow t}{\Gamma \vdash \langle \textbf{If } b\ c1\ c2,\ \text{Normal } s \rangle \Rightarrow t}\ (IfTrue) \qquad \frac{s \notin b \qquad \Gamma \vdash \langle c_2,\ \text{Normal } s \rangle \Rightarrow t}{\Gamma \vdash \langle \textbf{If } b\ c_1\ c_2,\ \text{Normal } s \rangle \Rightarrow t}\ (IfFalse)$$

$$\frac{s_1 \in b \qquad \Gamma \vdash \langle c,\ s_1 \rangle \Rightarrow s_2 \qquad \Gamma \vdash \langle \textbf{While } b\ c,\ s_2 \rangle \Rightarrow s_3}{\Gamma \vdash \langle \textbf{While } b\ c,\ \text{Normal } s_1 \rangle \Rightarrow s_3}\ (WhileTrue)$$

$$\frac{s \notin b}{\Gamma \vdash \langle \textbf{While } b\ c,\ \text{Normal } s \rangle \Rightarrow \text{Normal } s}\ (WhileFalse)$$

$$\frac{\Gamma \vdash \langle c\ s,\ \text{Normal } s \rangle \Rightarrow t}{\Gamma \vdash \langle \textbf{DynCom } c,\ \text{Normal } s \rangle \Rightarrow t}\ (DynCom) \qquad \frac{\Gamma \vdash \langle the\ (\Gamma\ f),\ \text{Normal } s \rangle \Rightarrow t}{\Gamma \vdash \langle \textbf{Call } f,\ \text{Normal } s \rangle \Rightarrow t}\ (Call)$$

$$\frac{\Gamma \vdash \langle c_1, \text{Normal } s \rangle \Rightarrow \text{Normal } t}{\Gamma \vdash \langle \textbf{Handle } c_1\ c_2, \text{Normal } s \rangle \Rightarrow \text{Normal } t}\ (HandleNormal)$$

$$\frac{\Gamma \vdash \langle c_1, \text{Normal } s \rangle \Rightarrow \text{Rev } r \qquad \Gamma \vdash \langle c_2,\ \text{Normal } r \rangle \Rightarrow t}{\Gamma \vdash \langle \textbf{Handle } c_1\ c_2,\ \text{Normal } s \rangle \Rightarrow t}\ (HandleRevert)$$

$$\frac{}{\Gamma \vdash \langle \textbf{Revert},\ \text{Normal } s \rangle \Rightarrow \text{Rev } s}\ (Revert) \qquad \frac{}{\Gamma \vdash \langle c,\ \text{Rev } s \rangle \Rightarrow \text{Rev } s}\ (RevState)$$

$$\frac{s \in b \qquad \Gamma \vdash \langle \textbf{Skip},\ Normal\ s \rangle \Rightarrow Normal\ s}{\Gamma \vdash \langle \textbf{Require } b,\ Normal\ s \rangle \Rightarrow Normal\ s}\ (RequireTrue)$$

$$\frac{s \notin b \qquad \Gamma \vdash \langle \textbf{Revert},\ Normal\ s \rangle \Rightarrow Rev\ s}{\Gamma \vdash \langle \textbf{Require } b,\ Normal\ s \rangle \Rightarrow Rev\ s}\ (RequireFalse)$$

$$\frac{\Gamma \vdash \langle bdy,\ Normal\ s \rangle \Rightarrow Normal\ t}{\Gamma \vdash \langle \textbf{Init } bdy\ rvrt,\ Normal\ s \rangle \Rightarrow Normal\ t}\ (ExecNormal)$$

$$\frac{\Gamma \vdash \langle bdy,\ Normal\ s \rangle \Rightarrow Rev\ t}{\Gamma \vdash \langle \textbf{Init } bdy\ rvrt,\ Normal\ s \rangle \Rightarrow Rev\ (rvrt\ s\ t)}\ (ExecRev)$$

Fig. 1. Big-step semantics rules for SOLI

Exceptions

To deal with exceptions, the EVM has two available opcodes: REVERT and INVALID. Both undo all state changes, but REVERT will also allow to return a value and refund all remaining gas to the caller, whereas INVALID will simply consume all remaining gas. Solidity uses these opcodes to handle exceptions using the revert(), require() and assert() functions. The require() and assert() functions receive a bool and throw the respective exception if the condition is not met while revert() simply throws the exception. Both revert()

and `require()` use the `REVERT` opcode and can also receive an error message to display to the user; `assert()` uses the `INVALID` opcode.

In `SOLI` the first two are modeled. `revert()` corresponds to the *Revert* command, which modifies the current state type from *Normal* to *Rev*. The `require()` function is defined by the *Require* command, as a conditional statement.

$$Require ::\; 's\; bexp \;\Rightarrow\; 's\; com$$
$$Require\; b \;:=\; \textbf{If } b \textbf{ Skip Revert}$$

Calling a Function

In order to model the different types of function call, calling must be extended with the following definition, which introduces the modeling of passing arguments, resetting local variables and returning results.

$$call ::\; [\![\; 's \Rightarrow' s,\; fname,\; 's \Rightarrow' s \Rightarrow' s,\; 's \Rightarrow' s \Rightarrow' s\; com\;]\!] \;\Rightarrow\; 's\; com$$
$$call\; pass\; f\; return\; result\; :=$$
$$\textbf{DynCom } (\lambda s.\; (\textbf{Upd } pass;;\; \textbf{Call } f;;$$
$$(\textbf{DynCom } (\lambda t.\; \textbf{Upd } (return\; s);;\; result\; s\; t))))$$

Here **DynCom** is used to abstract over the state space and refer to certain program states. The initial state s is captured by the first **DynCom** and the state after executing the body of the called procedure, t, by the second. The function *pass*, receives the initial state s and is used to pass the arguments of the function to the intended variables in the memory before the body of the function is executed. The *return* function is used to return from the procedure by cleaning the state, that is by restoring the local variables. In the case of a function call with a return value, the *result* function is used to communicate the results to the caller environment by updating the result variable. The control flow of *call* is depicted is in Fig. 2.

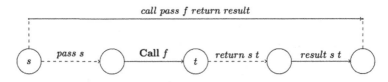

Fig. 2. Control flow for *call*

The big-step execution rule for *call*, Fig. 3, follows intuitively from the above description. First the body of the called function is executed after passing the arguments, that is, starting in state *pass s*. Then the result command is executed after returning from the call, that is, starting in state *return s t*.

$$\frac{(the \; (\Gamma f), \; Normal \; (pass \; s)) \; \Rightarrow \; Normal \; t \quad (result \; s \; t, \; Normal \; (return \; s \; t)) \; \Rightarrow \; u}{(call \; pass \; f \; return \; result, \; Normal \; s) \; \Rightarrow \; u} \; (ExecCall)$$

Fig. 3. Big-step execution rule for *call*

In Solidity, a function call can be internal or external. For *internal calls*, functions are in the same contract and so state variables, memory, and execution context are the same and we only need to model function arguments and results. *External calls*, which call functions from other contracts, are done via message call. All function arguments have to be copied to memory and, after execution, the memory needs to be restored. In addition, some execution environment variables are updated, such as msg_sender, msg_value, msg_data and $address\_this$.

Reverting State Changes

In Solidity, whenever an error occurs, for instance when some condition is not satisfied, a REVERT exception is thrown and all state changes made in the current call must be reverted.

Suppose one wants to execute the SOLI statement bdy starting in a normal type state s. The execution can run without any errors and terminate in a normal state t. But, if an exception is thrown, the execution must be stopped with the current Rev state t' in order to proceed to the state reversion. This is modeled with **Handle** bdy c, where c is the statement which handles the reversion.

Inside c the state is first passed to a normal state in order to allow the regular SOLI statements, for instance **Upd** to be executed. The update of the state variables to their original value is made with the $rvrt$ function, which receives the initial state s and the current state t. Finally the error is propagated by re-throwing **Revert**. The control flow for a statement execution is depicted in Fig. 4.

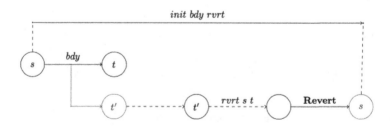

Fig. 4. Control flow for *Init*

In order to actually revert the state, first one needs to get hold of the initial state s which can be captured using **DynCom**. Also, while taking care of the state reversion, another **DynCom** is used to refer to the current state t when updating the variables.

Whenever a statement, such as a function, is written in SOLI it is encapsulated in an *Init* command which receives the function body and the *rvrt* function which models the reset of all state variables in case of error.

$$Init \ :: \ [\![\ 's \ com, \ 's \ \Rightarrow \ 's \ \Rightarrow \ 's]\!] \ \Rightarrow \ 's \ com$$
$$Init \ bdy \ rvrt \ := \ \textbf{DynCom} \ (\lambda s. \ (\textbf{Handle} \ bdy;; \ (\textbf{DynCom} \ (\lambda t.$$
$$\textbf{Upd} \ (rvrt \ s);; \ \textbf{Revert}))))$$

The big-step execution rules for *Init* are defined in Fig. 1. For normal execution it is immediate, it is just the regular execution of *bdy*. If an exception is thrown when executing *bdy*, its execution stops in a revert type state t and the full execution will terminate in state *rvrt s t*.

3.5   State Space

Since the goal of this work is to verify properties about specific programs, it was chosen to explicitly state the HOL type for each variable by working with records. Some of the used types are constructed using the HOL type word, which represents a bit.

$$byte \ := \ 8 \ word$$
$$address \ := \ 160 \ word$$
$$uint \ := \ 256 \ word$$

A record of Isabelle/HOL is a collection of fields where each has a specified name and type. A record comes with select and update operations for each field. Record types can also be defined by extending other record types. A record *st* represents the storage of a Solidity contract and *loc* the local variables for the functions in that contract. We illustrate this concept for an electronic voting contract in Sect. 5.1.

3.6   Environment Variables

Solidity defines a set of *global variables* regarding the execution environment, mainly to provide information about the blockchain. For a block, we need variables that keep track of the current block's hash, miner's address, difficulty, gaslimit, number and timestamp. For a transaction, we need variables that keep track of the current message call: data, gas, sender and signature, and for the whole transaction: gasprice and origin. In SOLI these variables are part of the environment record *env*, which is defined in Fig. 5.

An *account state* is defined as a record *Account* with four fields corresponding to its nonce, balance, storage and code, also represented in Fig. 5. The *world state* is defined as a field of the *env* record *gs* which maps addresses to their account states.

$$
\begin{array}{rcl}
record\ account\_state & := & \\
nonce & := & uint \\
balance & :: & uint \\
storage & :: & uint \Rightarrow uint \\
code & :: & byte\ list
\end{array}
\qquad
\begin{array}{rcl}
record\ env & := & \\
block\_coinbase & :: & uint \\
block\_difficulty & :: & address \\
block\_gaslimit & :: & uint \\
block\_number & :: & uint \\
block\_timestamp & :: & uint \\
msg\_data & :: & byte\ list \\
msg\_sender & :: & address \\
msg\_value & :: & uint \\
tx\_origin & :: & address \\
tx\_gasPrice & :: & uint \\
address\_this & :: & address \\
gs & :: & address \Rightarrow account\_state
\end{array}
$$

Fig. 5. Account state representation and environment variables

4   Hoare Logic

In this section we present Hoare logic, a system proposed by Tony Hoare [12,13], and its formalization for SOLI regarding partial correctness. We extended the proof system in [23] to a relevant subset of Solidity.

4.1   The Proof System

A Hoare logic formula is a triple of the form $P\ c\ Q$, where c is a command and P and Q are assertions, the precondition and postcondition, respectively. In EVM, command execution can result in a *Normal* or in a *Rev* state. To model this feature, we split the postcondition in two, Q and A, for regular and for exceptional termination respectively.

To reason about recursive procedures, a set of assumptions Θ is introduced. This set contains function specifications, which will be used as hypothesis when proving the body of a recursive procedure. An *assumption for a function* is a tuple that contains its precondition, name and both postconditions.

$$'s\ assmpt := \langle 's\ assn,\ fname,\ 's\ assn,\ 's\ assn \rangle$$

The notation used for a derivable Hoare formula is associated with the procedure body environment Γ and with the set of assumptions Θ.

Definition 8 (Hoare logic). *A Hoare logic is defined for SOLI such that a derivable formula is represented by*

$$\Gamma, \Theta \vdash P\ c\ Q, A$$

$$\frac{}{\Gamma, \Theta \vdash Q \textbf{ Skip } Q, A} \; (Skip) \qquad \frac{}{\Gamma, \Theta \vdash \{s.\; f\; s \in Q\} \;(\textbf{Upd } f)\; Q, A} \;(Upd)$$

$$\frac{\Gamma, \Theta \vdash P\; c_1\; R, A \quad \Gamma, \Theta \vdash R\; c_2\; Q, A}{\Gamma, \Theta \vdash P\;(\textbf{Seq } c_1\; c_2)\; Q, A} \;(Seq) \qquad \frac{\Gamma, \Theta \vdash (P \cap b)\; c\; P, A}{\Gamma, \Theta \vdash P\;(\textbf{While } b\; c)\;(P \cap {-b}), A} \;(While)$$

$$\frac{\Gamma, \Theta \vdash (P \cap b)\; c_1\; Q, A \quad \Gamma, \Theta \vdash (P \cap {-b})\; c_2\; Q, A}{\Gamma, \Theta \vdash P\;(\textbf{If } b\; c_1\; c_2)\; Q, A} \;(If)$$

$$\frac{\Gamma, \Theta \vdash P\; c_1\; Q, R \quad \Gamma, \Theta \vdash R\; c_2\; Q, A}{\Gamma, \Theta \vdash P\;(\textbf{Handle } c_1\; c_2)\; Q, A} \;(Handle) \qquad \frac{}{\Gamma, \Theta \vdash A \textbf{ Revert } Q, A} \;(Revert)$$

$$\frac{\forall s \in P.\;\; \Gamma, \Theta \vdash P\;(c\; s)\; Q, A}{\Gamma, \Theta \vdash P\;(\textbf{DynCom } c)\; Q, A} \;(DynCom)$$

$$\frac{(P, f, Q, A) \in S \quad \forall \langle P, f, Q, A \rangle \in S.\; f \in dom\; \Gamma \wedge \Gamma, (\Theta \cup S) \vdash P\;(the\;(\Gamma\; f))\; Q, A}{\Gamma, \Theta \vdash P\;(\textbf{Call } f)\; Q, A} \;(CallRec)$$

$$\frac{\Gamma, \Theta \vdash (P \cap b)\textbf{ Skip } Q, A \quad \Gamma, \Theta \vdash (P \cap {-b})\textbf{ Revert } Q, A}{\Gamma, \Theta \vdash P\;(\textbf{Require } b)\; Q, A} \;(Require)$$

$$\frac{\forall s \in P.\;\; \Gamma, \Theta \vdash P\; bdy\; Q, \{t.\; rvrt\; s\; t \in A\}}{\Gamma, \Theta \vdash P\;(\textbf{Init } bdy\; rvrt)\; Q, A} \;(Init) \qquad \frac{(P, f, Q, A) \in \Theta}{\Gamma, \Theta \vdash P\;(\textbf{Call } f)\; Q, A} \;(Asm)$$

$$\frac{\forall s \in P'.\; \exists P Q A.\; \Gamma, \Theta \vdash P\; c\; Q, A \wedge Q \subseteq Q' \wedge A \subseteq A' \wedge s \in P}{\Gamma, \Theta \vdash P'\; c\; Q', A'} \;(Conseq)$$

Fig. 6. Hoare logic for *SOLI*

where

$$\Gamma :: fname \Rightarrow {'s}\; com$$
$$\Theta :: {'s}\; assmpt\; set$$
$$P, Q, A :: {'s}\; assn$$
$$c :: {'s}\; com,$$

and the proof system is constituted by the rules in Fig. 6.

There is a rule for each SOLI command and, additionally, the *Asm* and *Consequence* rules. To have an intuitive meaning for the rules of this system, one should read it backwards. For instance, for the *Upd* rule, if Q holds (for regular execution) after the update then P is the set of states such that the application of f to them belongs to Q. *Skip* and *Revert* have the intuitive meaning of doing nothing, and *Seq* and *Handle* correspond respectively to the cases where $c_1\; c_2$ are both executed, and c_1 throws an exception and c_2 is executed. In the *DynCom* rule, the triple has to hold for every state s that satisfies the precondition as the dynamic command will depend on the initial state. The *CallRec* rule regards a set of function specifications S whose bodies are verified and that is added to Θ. Then, when one of these functions is called, the specification can be assumed using *Asm* rule.

The **Require** command is modelled as a conditional statement, hence both rules follow the same structure. In one of the branches the precondition b holds, and in the other it does not. The **Init** statement corresponds to a regular execution of the body ending in a state for which the regular postcondition Q holds or, in the case of an exception being thrown, the execution ends in a state such that by reverting all state changes the exceptional postcondition A holds. A dependence on the initial state s is introduced in the premise. The *Consequence* rule allows to strengthen the precondition as well as to weaken the postcondition.

4.2   Weakest Precondition Calculus

In order to verify properties about programs using Hoare logic, a backward propagation method is followed. In this method, sufficient conditions for a certain result, the postcondition, are determined. The rules are successively applied backwards, starting in the postcondition until the beginning of the program. Some side conditions may be generated.

The weakest precondition is then said to be the most lenient assumption on the initial state such that Q, A will hold after the execution of the command c.

Weakest precondition calculus, also know as predicate transformer semantics (Dijkstra [6]), is a reformulation of Hoare logic. It constitutes a strategy to reduce the problem of proving a Hoare formula to the problem of proving an HOL assertion, which is called the verification condition. Since assertions are expressed as sets, reasoning about the conditions is expressed using set operations.

Definition 9 (Weakest precondition calculus). *Let c be a command, Q and A assertions, and $wp^{\Gamma,\Theta}(c, Q, A)$ the weakest precondition of Q, A for c. The weakest precondition calculus for SOLI is inductively defined as follows:*

$wp^{\Gamma,\Theta}$ (**Skip**, Q, A) $= Q$

$wp^{\Gamma,\Theta}$ (**Revert**, Q, A) $= A$

$wp^{\Gamma,\Theta}$ (**Upd** f, Q, A) $= \{s.\ f\ s \in Q\}$

$wp^{\Gamma,\Theta}$ (**Seq** $c_1\ c_2$, Q, A) $= wp^{\Gamma,\Theta}\ (c_1,\ wp^{\Gamma,\Theta}\ (c_2,\ Q,\ A),\ A)$

$wp^{\Gamma,\Theta}$ (**If** $b\ c_1\ c_2$, Q, A) $= \{s.\ (s \in b \longrightarrow s \in wp^{\Gamma,\Theta}\ (c_1,\ Q,\ A)) \wedge$
$$(s \notin b \longrightarrow s \in wp^{\Gamma,\Theta}\ (c_2,\ Q,\ A))\}$$

$wp^{\Gamma,\Theta}$ (**While** $I\ b\ c$, Q, A) $=$
$$\{s.\ (s \in b \longrightarrow s \in wp^{\Gamma,\Theta}\ (\textbf{Seq}\ c\ (\textbf{While}\ I\ b\ c),\ Q,\ A)) \wedge$$
$$(s \notin b \longrightarrow s \in wp^{\Gamma,\Theta}\ (c_2,\ Q,\ A))\}$$

$wp^{\Gamma,\Theta}$ (**Call** f, Q, A) $= P_f$, such that $f \in dom\ \Gamma \wedge (P_f,\ f,\ Q_f,\ A_f) \in \Theta$

$wp^{\Gamma,\Theta}$ (**DynCom** c, Q, A) $= \bigcap_{s} wp^{\Gamma,\Theta}\ (c\ s,\ Q,\ A)$

$wp^{\Gamma,\Theta}$ (**Handle** $c_1\ c_2$, Q, A) $= wp^{\Gamma,\Theta}\ (c_1,\ Q,\ wp^{\Gamma,\Theta}\ (c_2,\ Q,\ A))$

$wp^{\Gamma,\Theta}$ (**Require** b, Q, A) $= \{s.\ (s \in b \longrightarrow s \in Q \wedge (s \notin b \longrightarrow s \in A)\}$

$wp^{\Gamma,\Theta}$ (**Init** $bdy\ rvrt$, Q, A) $= \bigcap_{s} wp^{\Gamma,\Theta}\ (bdy,\ Q,\ \{t.\ rvrt\ s\ t \in A\})$

The weakest precondition for the call of procedure f corresponds to the precondition for its specification, present in the set of assumptions.

Since both **Dyncom** and **Init** have to consider every preceding state s, their weakest precondition corresponds to the intersection of certain weakest preconditions: in the former, to the wp of the command applied to each one of the states; in the latter, to the wp of bdy such that in case of exception the state reversion is applied to s. From the definition of **Require** it follows immediately that its weakest precondition is Q if b holds, and A otherwise.

A strategy to generate verification conditions based on this calculus is described in Sect. 4.5.

4.3 Soundness

To prove soundness of our proof system, we follow the same technique as [23]. The formal definition for validity regarding partial correctness is defined as follows:

Definition 10 (Validity—Partial Correctness)

$$\Gamma \vDash P\ c\ Q, A \quad if$$
$$\forall s\ t.\ \Gamma \vdash \langle c, s \rangle \Rightarrow t \wedge s \in Normal\ 'P \longrightarrow t \in Normal\ 'Q \cup Rev\ 'A.$$

The goal is to prove that if a formula is derivable in the Hoare Logic (Fig. 6) then it also valid according to Definition 10. In the case of recursive calls, we need to take into account the set of assumptions Θ and also the depth of recursion. However, the definitions of validity and big-step semantics are not rich enough to approach these properties. The notion of validity is thus extended with the set of assumptions.

Definition 11 (Validity with context)

$$\Gamma, \Theta \vDash P\ c\ Q, A \quad if$$
$$\forall \langle P, f, Q, A \rangle \in \Theta.\ \Gamma \vDash P\ (\textbf{Call } f)\ Q, A \quad \longrightarrow \quad \Gamma \vDash P\ c\ Q, A.$$

Also, an additional set of big-step rules to deal with the depth of recursion are defined, where n is the limit on nested procedure calls.

$$\Gamma \vdash \langle c,\ s \rangle \overset{n}{\Rightarrow} t$$

These rules are similar to the normal big-step rules (Fig. 1) except for the *Call* statement where the limit n is decremented in each step to account for the depth of the recursion.

$$\frac{\Gamma \vdash \langle the\ (\Gamma\ f),\ \text{Normal } s \rangle \overset{n}{\Rightarrow} t}{\Gamma \vdash \langle \textbf{Call } f,\ \text{Normal } s \rangle \overset{n+1}{\Rightarrow} t}\ (Call)$$

We can show that this new set of rules is monotonic with respect to the limit n.

Lemma 1 (Monotonicity)

$$\Gamma \vdash \langle c, s \rangle \overset{n}{\Rightarrow} t \wedge n \leq m \quad \longrightarrow \quad \Gamma \vdash \langle c, s \rangle \overset{m}{\Rightarrow} t$$

Validity can now be established regarding the limit on nested recursive calls.

Definition 12 (Validity with limit)

$$\Gamma \overset{\shortmid}{\vDash} P\ c\ Q, A \quad if$$
$$\forall s\ t.\ \Gamma \vdash \langle c, s \rangle \overset{n}{\Rightarrow} t \wedge s \in Normal\ 'P \longrightarrow t \in Normal\ 'Q\ \cup\ Rev\ 'A.$$

Finally the notions of validity with context and limit can be joined, leading to a definition which suits the needs to reason about recursive procedure calls.

Definition 13 (Validity with limit and context)

$$\Gamma, \Theta \overset{\shortmid}{\vDash} P\ c\ Q, A \quad if$$
$$\forall \langle P, f, Q, A \rangle \in \Theta.\Gamma \overset{\shortmid}{\vDash} P\ (\textbf{Call } f)\ Q, A \quad \longrightarrow \quad \Gamma \overset{\shortmid}{\vDash} P\ c\ Q, A.$$

The required conditions to show that Hoare rules preserve validity are now established.

Lemma 2

$$(\forall n.\ \Gamma, \Theta \overset{n}{\vDash} P\ c\ Q, A) \ \longrightarrow\ \Gamma, \Theta \vDash P\ c\ Q, A$$

Lemma 3 (Soundness with limit and context). *Let Γ be the mapping between function names and their bodies, Θ the set of assumptions, c a SOLI command, P the precondition and Q, A the postconditions.*

$$If\ \Gamma, \Theta \vdash P\ c\ Q, A\ then\ (\forall n.\ \Gamma, \Theta \overset{n}{\vDash} P\ c\ Q, A).$$

The intended result follows directly from Lemmas 3 and 2.

Theorem 1 (Soundness). *Let Γ be the mapping between function names and their bodies, Θ the set of assumptions, c a SOLI command, P the precondition and Q, A the postconditions.*

$$If\ \Gamma, \Theta \vdash P\ c\ Q, A\ then\ \Gamma, \Theta \vDash P\ c\ Q, A.$$

4.4   Completeness

Due to its inheritance from HOL, used to state assertions, Hoare logic is not complete. However, Cook [5] introduced the notion of relative completeness by separating incompleteness of the assertion language from incompleteness due to inadequacies in the axioms and rules for the programming language constructs. It is assumed that there is an oracle which can be inquired about the validity of an HOL assertion. The proof follows the method by Winskel [25] and relies on the concept of weakest precondition, Definition 9.

Lemma 4

$$\Gamma, \Theta \vDash P\ c\ Q, A \ \longrightarrow\ (s \in P \ \longrightarrow\ s \in wp(c,\ Q,\ A))$$

An auxiliary weakest precondition property regarding the derivation of a formula and its precondition is proven.

Lemma 5

$$\Gamma, \Theta \vdash\ wp(c, Q, A)\ c\ Q, A$$

Using Lemma 5, the (relative) completeness Theorem can now be proven.

Theorem 2 ((Relative) Completeness). *Let Γ be the mapping between function names and their bodies, Θ the set of assumptions, c a SOLI command, P the precondition and Q, A the postconditions.*

$$If\ \Gamma, \Theta \vDash P\ c\ Q, A\ then\ \Gamma, \Theta \vdash P\ c\ Q, A.$$

4.5   Computation of Verification Conditions

One of the goals of this work is to develop a proof technique for the verification of properties about smart contracts. In this section, we extend previous work by Frade and Pinto [7] and present a method to compute the verification conditions of a program, which follows a backwards propagation through the weakest precondition.

To achieve this, the invariants for while loops must be supplied explicitly. The concept of annotated command is, therefore, introduced.

Definition 14 (Annotated commands). *The syntax for* annotated commands *is defined by the polymorphic datatype* $'s$ *acom.*

$$
\begin{aligned}
's\ acom\ :=\ &\textbf{Skip}\mid \textbf{Upd}\ 's\ \Rightarrow\ 's\ \mid\ \textbf{Seq}\ 's\ acom\ 's\ acom \\
&\mid\ \textbf{If}\ 's\ bexp\ 's\ acom\ 's\ acom\ \mid\ \textbf{While}\ 's\ assn\ 's\ bexp\ 's\ acom \\
&\mid\ \textbf{Dyncom}\ 's\ \Rightarrow\ 's\ acom\ \mid\ \textbf{Call}\ fname\ \mid\ \textbf{Revert} \\
&\mid\ \textbf{Handle}\ 's\ acom\ 's\ acom\ \mid\ \textbf{Require}\ 's\ bexp \\
&\mid\ \textbf{Init}\ 's\ acom\ 's\ \Rightarrow\ 's\ \Rightarrow\ 's
\end{aligned}
$$

The weakest precondition calculus for annotated commands is the same as for normal commands except for **While** where it becomes the loop invariant, since it is a condition that must be met before each loop execution, or even if it isn't executed in the first place.

Definition 15 (Weakest precondition calculus for annotated commands). *The weakest precondition calculus for annotated commands is inductively defined as follows:*

$$wp^{\Gamma,\Theta}\ (\textbf{Skip},\ Q,\ A)\ =\ Q$$

$$wp^{\Gamma,\Theta}\ (\textbf{Revert},\ Q,\ A)\ =\ A$$

$$wp^{\Gamma,\Theta}\ (\textbf{Upd}\ f,\ Q,\ A)\ =\ \{s.\ f\ s\in Q\}$$

$$wp^{\Gamma,\Theta}\ (\textbf{Seq}\ c_1\ c_2,\ Q,\ A)\ =\ wp^{\Gamma,\Theta}\ (c_1,\ wp^{\Gamma,\Theta}\ (c_2,\ Q,\ A),\ A)$$

$$
\begin{aligned}
wp^{\Gamma,\Theta}\ (\textbf{If}\ b\ c_1\ c_2,\ Q,\ A)\ =\ \{s.\ (s\in b\ &\longrightarrow\ s\in wp^{\Gamma,\Theta}\ (c_1,\ Q,\ A))\ \wedge \\
(s\notin b\ &\longrightarrow\ s\in wp^{\Gamma,\Theta}\ (c_2,\ Q,\ A))\}
\end{aligned}
$$

$$wp^{\Gamma,\Theta}\ (\textbf{While}\ I\ b\ c,\ Q,\ A)\ =\ I$$

$$wp^{\Gamma,\Theta}\ (\textbf{Call}\ f,\ Q,\ A)\ =\ P_f\ \text{such that}\ f\in dom\ \Gamma\ \wedge\ (P_f,\ f,\ Q_f,\ A_f)\in\Theta$$

$$wp^{\Gamma,\Theta}\ (\textbf{DynCom}\ c,\ Q,\ A)\ =\ \bigcap_s\ wp^{\Gamma,\Theta}\ (c\ s,\ Q,\ A)$$

$$wp^{\Gamma,\Theta}\ (\textbf{Handle}\ c_1\ c_2,\ Q,\ A)\ =\ wp^{\Gamma,\Theta}\ (c_1,\ Q,\ wp^{\Gamma,\Theta}\ (c_2,\ Q,\ A))$$

$$wp^{\Gamma,\Theta}\ (\textbf{Require}\ b,\ Q,\ A)\ =\ \{s.\ (s\in b\ \longrightarrow\ s\in Q\ \wedge\ (s\notin b\ \longrightarrow\ s\in A)\}$$

$$wp^{\Gamma,\Theta}\ (\textbf{Init}\ bdy\ rvrt,\ Q,\ A)\ =\ \bigcap_s\ wp^{\Gamma,\Theta}\ (bdy,\ Q,\ \{t.\ rvrt\ s\ t\in A\})$$

The verification condition computation for a command can be obtained using the structure of each rule in the proof system. An important property about these is that the verification conditions are computed independently from preconditions, leaving only the need to check their inclusion in the propagated weakest precondition. This prevents the generation of unnecessary verification conditions.

Definition 16 (Verification condition I). *The* verification condition *function vc is defined as follows:*

$$vc\ (\Gamma, \Theta \vdash P\ c\ Q, A) \ = \ P \subseteq wp^{\Gamma, \Theta}\ (c,\ Q,\ A)\ \cup\ vc_{aux}^{\Gamma, \Theta}\ (c,\ Q,\ A),$$

where the auxiliary verification condition $vc_{aux}^{\Gamma, \Theta}$ is inductively defined as follows:

$vc_{aux}^{\Gamma, \Theta}$ (**Skip**, Q, A) $= \emptyset$

$vc_{aux}^{\Gamma, \Theta}$ (**Revert**, Q, A) $= \emptyset$

$vc_{aux}^{\Gamma, \Theta}$ (**Upd** f, Q, A) $= \emptyset$

$vc_{aux}^{\Gamma, \Theta}$ (**Seq** c_1 c_2, Q, A) $= vc_{aux}^{\Gamma, \Theta}\ (c_1,\ wp^{\Gamma, \Theta}(c_2,\ Q,\ A),\ A)\ \cup\ vc_{aux}^{\Gamma, \Theta}\ (c_2,\ Q,\ A)$

$vc_{aux}^{\Gamma, \Theta}$ (**If** b c_1 c_2, Q, A) $= vc_{aux}^{\Gamma, \Theta}\ (c_1,\ Q,\ A)\ \cup\ vc_{aux}^{\Gamma, \Theta}\ (c_2,\ Q,\ A)$

$vc_{aux}^{\Gamma, \Theta}$ (**While** I b c, Q, A) $= (I \cap b) \subseteq wp^{\Gamma, \Theta}\ (c,\ I,\ A)\ \cup$

$$vc_{aux}^{\Gamma, \Theta}\ (c,\ I,\ A)\ \cup\ (I \cap -b) \subseteq Q$$

$vc_{aux}^{\Gamma, \Theta}$ (**Call** f, Q, A) $= Q_f \subseteq Q$

$vc_{aux}^{\Gamma, \Theta}$ (**DynCom** c, Q, A) $= \bigcap_s vc_{aux}^{\Gamma, \Theta}\ (c\ s,\ Q,\ A)$

$vc_{aux}^{\Gamma, \Theta}$ (**Handle** c_1 c_2, Q, A) $= vc_{aux}^{\Gamma, \Theta}\ (c_1,\ Q,\ wp^{\Gamma, \Theta}(c_2,\ Q,\ A))\ \cup vc_{aux}^{\Gamma, \Theta}\ (c_2,\ Q,\ A)$

$vc_{aux}^{\Gamma, \Theta}$ (**Require** b, Q, A) $= \emptyset$

$vc_{aux}^{\Gamma, \Theta}$ (**Init** bdy $rvrt$, Q, A) $= \bigcap_s vc_{aux}^{\Gamma, \Theta}\ (bdy,\ Q,\ \{t.\ rvrt\ s\ t \in A\})$

However, upon the verification of a program with any number of function calls, their specification must have been verified and added to the set of assumptions. A verification condition suitable for every program is then formalized.

Definition 17 (Verification condition II). *Let S be the set of specifications for every function whose call is generated by the execution of c. The verification condition function VC for c is defined as*

$$VC\ (\Gamma, \Theta \vdash P\ c\ Q, A) \ = \ P \subseteq wp^{\Gamma, \Theta}\ (c,\ Q,\ A)\ \cup$$
$$vc_{aux}^{\Gamma, \Theta}\ (c,\ Q,\ A)\ \cup\ \bigcup_{\langle P, f, Q, A \rangle \in S}\ vc\ (\Gamma, (\Theta \cup S) \vdash P\ (the\ (\Gamma\ f))\ Q, A).$$

An Alternative Formulation of Rules

The verification condition computations explicitly separate the main verification condition (the inclusion of precondition in the weakest precondition of the program) from auxiliary conditions (generated from the structure of the rules). In order to construct a proof which follows this backwards propagation method, some Hoare rules are modified to a structure that will be referred as weakest precondition style. Following the method above, we were able to obtain the same rules as in [23] together with a new rule for the Solidity command *Require*. The set of rules is presented in Fig. 7.

$$\frac{P \subseteq Q}{\Gamma, \Theta \vdash P \ \mathbf{Skip} \ Q, A} \ (Skip') \qquad \frac{P \subseteq A}{\Gamma, \Theta \vdash P \ \mathbf{Revert} \ Q, A} \ (Revert')$$

$$\frac{P \subseteq \{s. \ (s \in b \ \longrightarrow \ s \in P_1) \ \wedge \ (s \notin b \ \longrightarrow \ s \in P_2)\} \qquad \Gamma, \Theta \vdash P_1 \ c_1 \ Q, A \qquad \Gamma, \Theta \vdash P_2 \ c_2 \ Q, A}{\Gamma, \Theta \vdash P \ (\mathbf{If} \ b \ c_1 \ c_2) \ Q, A} \ (If')$$

$$\frac{P \subseteq \{s. \ f \ s \ \in Q\}}{\Gamma, \Theta \vdash P \ (\mathbf{Upd} \ f) \ Q, A} \ (Upd')$$

$$\frac{P \subseteq I \qquad \Gamma, \Theta \vdash (I \cap b) \ c \ I, A \qquad (I \cap -b) \ \subseteq Q}{\Gamma, \Theta \vdash P \ (\mathbf{While} \ b \ c) \ Q, A} \ (While')$$

$$\frac{P \subseteq \{s. \ (s \in b \ \longrightarrow \ s \in Q) \ \wedge \ (s \notin b \ \longrightarrow \ s \in A)\}}{\Gamma, \Theta \vdash P \ (Require \ b) \ Q, A} \ (Require')$$

$$\frac{P \subseteq \{s. \ \exists Z. \ pass \ s \in P' \ Z \ \wedge \ (\forall t. \ t \in Q' \ Z \ \longrightarrow \ return \ s \ t \in R \ s \ t)\} \qquad \forall Z. \ \Gamma, \Theta \vdash (P' \ Z) \ (Call \ f) \ (Q' \ Z), A \qquad \forall st. \ \Gamma, \Theta \vdash (R \ s \ t) \ (result \ s \ t) \ Q, A}{\Gamma, \Theta \vdash P \ (call \ pass \ f \ return \ result) \ Q, A} \ (CallRec')$$

Fig. 7. Weakest precondition style rules

5   Application to Real-World Smart Contracts

In this section we illustrate the usage of our method for proving properties about smart contracts.

5.1   Electronic Voting

In this example an electronic voting contract, *Ballot*[1], which features automatic and transparent vote counting and delegate voting, is presented. This is an example of a successful contract verification that has some complex properties originated by the loop invariant, and that introduces the need to prove additional lemmas, defined generally.

The *Ballot* contract contains a *Voter* struct constituted by the weight of the voter (accumulated by delegation), a boolean that states whether the person already voted, the delegate's address (in case of vote delegation) and the index of the voted proposal. It also contains a *Proposal* struct constituted by the proposal name and corresponding vote count. As global variables the contract contains

---

[1] https://solidity.readthedocs.io/en/v0.5.12/solidity-by-example.html.

an address *chairperson*, a mapping *voters* between addresses and *Voter* structs, and a list of proposals *proposals*, which are stored in the *st* record.

record $st = env +$		**record** $loc = st +$
chairperson :: *address*	*winningVoteCount* :: *int*	
voters :: *address* $\Rightarrow$ *Voter*	p :: *int*	
proposals :: *Proposal list*	*winningProposal_out* :: *int*	
	r :: *int*	
	winningVoteCount_out :: *int*	

This example is focused on the verification of the *winnerName* function (Fig. 8), which returns the name of the winning proposal by calling the *winningProposal* function which returns the corresponding index. This function finds the maximum value of *voteCount* in the list of proposals using a loop. It introduces the necessity of supplying an invariant and to verify that, while the list is gone through, the current maximum is correctly computed. The verification requires a definition of the maximum of a list and additional lemmas on the matter to be introduced.

$winningProposal\_com$ :: $loc\ com$
$winningProposal\_com \equiv INIT($
    $'winningProposal\_out ::= 0;;$
    $'winningVoteCount ::= voteCount\ 'proposals[0];;$
    $'p ::= 1;;$
    WHILE $('p < length\ 'proposals)$ **DO**
        IF $('winningVoteCount < voteCount\ 'proposals['p])$
        THEN $'winningVoteCount ::= voteCount\ 'proposals['p];;$
            $'winningProposal\_out ::= \ 'p;;$
        $'p ::=' p + 1)$

$winnerName\_com$ :: $loc\ com$
$winnerName\_com \equiv INIT($
    $call\_wp;;$
    $'winnerName\_out ::= name\ 'proposals['r])$

$call\_wp$ :: $st\ com$
$call\_wp \equiv call\ (\lambda s.\ s)\ winningProposal\ (\lambda st.\ t)$
    $(\lambda st.\ \textbf{Upd}\ (\lambda u.\ u(\!|r := winningProposal\_out\ t|\!)))$

Fig. 8. *winningProposal* and *winnerName* functions

INIT is defined as an *init* statement to revert all state changes in case of exception, that is, resetting the global variables to their initial values. In order to internally call the *winningProposal* function, *call_wp* is defined as a *call* statement.

The verification consists in showing that the return value r from the *winningProposal* function corresponds to the maximum vote count of the list and that the output of the *winnerName* function is the corresponding name. The initial values for global variables are stored in the auxiliary variables *chair*, *vtrs* and *prop*.

$$\Gamma, \Theta \vdash \; \{| \; chair =' chairperson \; \wedge \; vtrs =' voters \; \wedge \; prop =' proposals \; |\}$$
$$winnerName\_com$$
$$\{| \; max' \; (map \; voteCount \; prop) \; = \; (map \; voteCount \; prop)['r] \; \wedge$$
$$'winnerName\_out \; = \; name \; prop['r] \; |\},$$
$$\{| \; 'chairperson = chair \; \wedge \; 'voters = vtrs \; \wedge \; 'proposals = prop \; |\}$$

$$I = \{| \; 1 \; \leq \; 'p \; \leq \; length \; prop \; \wedge$$
$$'winningVoteCount = max'(take \; 'p(map \; voteCount \; prop)) \; \wedge$$
$$'winningVoteCount = (map \; voteCount \; prop)['winningProposal\_out] \; |\}$$

The *max'* function was defined to retrieve the maximum of a list of natural numbers. Invariant I states the limits that should be verified on the value of p and that the current maximum is correctly computed.

The application of the verification method results, after simplification, in two conditions, which are solved through the use of the auxiliary lemmas.

1. $'proposals \neq \{\} \implies 1 \leq length \; 'proposals \; \wedge$
   $voteCount \; 'proposals[0] = max'(take \; 1 \; (map \; voteCount \; 'proposals))$
2. $'p < length \; 'proposals \implies$
   $(max'(take \; 'p \; (map \; voteCount \; 'proposals)) < voteCount \; 'proposals['p] \longrightarrow$
   $max'(take('p+1)(map \; voteCount \; 'proposals)) = voteCount \; 'proposals['p]) \; \wedge$
   $(\neg max'(take \; 'p \; (map \; voteCount \; 'proposals)) < voteCount \; 'proposals['p] \longrightarrow$
   $max' \; (take \; ('p+1) \; (map \; voteCount \; 'proposals)) =$
   $max' \; (take \; 'p \; (map \; voteCount \; 'proposals)))$

The first condition results from the precondition inclusion and is proved using Lemma 6, together with the fact that

$$voteCount \; 'proposals[0] = (map \; voteCount \; 'proposals)[0]$$

Lemma 6
$$l \neq \{\} \implies l[0] = max' \; (take \; 1 \; l).$$

The second, which results from the invariant verification conditions, is proven using the max' definition and Lemma 7, that follows by induction on the structure of the list.

Lemma 7

$$\{x_1, \ldots, x_n\} \neq \{\} \implies max'\ (\{x_1, \ldots, x_n\}) = max\ (max'\ \{x_1, \ldots, x_{n-1}\})\ x_n$$

5.2   Ethereum Tokens

Solidity is prone to underflows and overflows since the EVM works with 256-bit unsigned integers and, therefore, all operations are performed modulo 2^{256}. As an example of a vulnerable implementation of an ERC20 token, the Hexagon (HXG) token[2] is taken into account. This example illustrates that some proofs on the alleged specification of a contract may not terminate but give us important insights about the source of its vulnerability. Amongst its global variables it contains a mapping $balanceOf$, a mapping $allowances$ and a $uint$ $burnPerTransaction$, which is set to 2. In this example we analyze the $transfer$ function (Fig. 9). According to its specification it should be the case that, after the function is executed, the balance of address $from$ decreases by $val + 2$, the balance of address to increases by val, and the balance of address $adr0$ increases by 2.

Note that the conditions in the postcondition are stated using the $uint$ Isabelle function which allows to check that an operation does not underflow or overflow. The $uint\_arith$ Isabelle tactic is used in the proof to unfold this definition, which can take some time to run. It gets stuck with a verification condition which depends on the fact that $uint\ (val + 2) = uint\ val + 2$.

```
transfer :: loc com
transfer ≡ INIT (
REQUIRE ('to ≠ 'adr0);;
REQUIRE ('balanceOf 'frm ≥ 'val + 'burnPerTransaction);;
REQUIRE ('balanceOf 'to + 'val ≥ 'balanceOf 'to);;
'balanceOf ::= 'balanceOf ('frm := 'balanceOf 'frm − ('val + 'burnPerTransaction));;
'balanceOf ::= 'balanceOf ('to := 'balanceOf 'to + 'val);;
'balanceOf ::= 'balanceOf ('adr0 := 'balanceOf 'adr0 + 'burnPerTransaction);;
'currentSupply ::= 'currentSupply − 'burnPerTransaction)
```

Fig. 9. *transfer* function of *Hexagon* contract

---

[2] https://etherscan.io/address/0xB5335e24d0aB29C190AB8C2B459238Da1153cEBA #code.

$$\Gamma, \Theta \vdash \{|\ 'burnPerTransaction = 2\ \wedge\ from = 'frm\ \wedge\ t = 'to\ \wedge\ a = 'adr0\ \wedge$$
$$bal\_from = 'balanceOf\ from\ \wedge\ bal\_to = 'balanceOf\ t\ \wedge$$
$$bal\_a = 'balanceOf\ a\ \wedge\ supply = 'currentSupply\ \wedge$$
$$from \neq a\ \wedge\ from \neq t\ \wedge\ a \neq t\ |\}$$

$transfer$

$$\{|\ uint('balanceOf\ from) = uint\ bal\_from\ -\ (uint\ 'val + 2)\ \wedge$$
$$uint('balanceOf\ t) = uint\ bal\_to + uint\ 'val\ \wedge$$
$$uint('balanceOf\ a) = uint\ bal\_a + 2\ |\},$$
$$\{|\ 'balanceOf\ from = bal\_from\ \wedge\ 'balanceOf\ t = bal\_to\ \wedge$$
$$'balanceOf\ a = bal\_a\ |\}$$

Looking at *transfer* function there is no condition that ensures this and prevents overflow during the addition of $'val$ and $'burnPerTransaction$. Therefore, one is not able to prove the specification since there is no way to prove that

$$uint\ (balanceOf\ from) = uint\ (bal\_from - (val + 2))$$
$$= uint\ bal\_from - (uint\ val + 2)$$

This vulnerability can be exploited. Suppose the *transfer* function is called by an attacker with *val* equal to $2^{256} - 2$. It follows that $val + burnPerTransaction = 2^{256} - 2 + 2 = 0$ and therefore the second **REQUIRE** statement's guard will become $balanceOf\ 'frm \geq 0$ which is always true. The balance of $'frm$ is then decreased by 0 and the balance of $'to$ increased by $2^{256} - 2$.

To solve this issue a require statement can be added to the *transfer* function which checks if $'val +' burnPerTransaction < 2^{256}$.

5.3   Reentrancy

This example shows how the defined recursive features can be used to model reentrancy. A *DAO* is a Decentralized Autonomous Organization built using the Ethereum blockchain. In 2016, an hacker exploited a bug in the *DAO* contract which resulted in the loss of approximately \$50 million in ether. This was the first reentrancy attack which consisted in draining funds using the attacker's fallback function.

A fallback function is a contract's function, with no arguments or return values, which is automatically executed whenever a call is made to the contract and none of its other functions match the given function identifier or when no data is supplied. This is the case when the contract receives ether, with no data specified. The vulnerability consisted in the fact that *DAO*'s *withdraw* function uses *call.value()* to send ether to the caller's account. Now, this triggers its fallback function, which contains arbitrary code defined by the owner.

To explain the technical aspects of this attack a simplified version, *babyDao*, is presented. The contract contains, as state variable, a mapping *credit* between addresses and their respective balances. The vulnerability is present in the

$withdraw\_com$ :: $loc\ com$

$withdraw\_com \equiv INIT($

   IF $('credit\ user > 0)$

   THEN $call\_value; ;$

   $'credit ::= 'credit(user := 0))$

$fallback$ :: $loc\ com$

$fallback \equiv INIT($

   IF $(balance\ ('gs\ babyDao) - 'credit\ user \geq 0)$

   THEN $call\_withdraw)$

Fig. 10. *withdraw* and malicious *fallback* function

withdraw function (Fig. 10) and the attacker's goal is to drain all caller's funds in the contract and send this value to his account. To perform this transference the function uses *call.value()* which triggers its fallback function, containing arbitrary code defined by the owner. This is modelled as the *call* statement *call_value*, which is defined so that the values of some environment variables are updated, the balance of *user* is increased by *msg_value* and the balance of *babyDao* is decreased by the same amount. The fallback function's code is then executed.

To write the specification for *withdraw*, the auxiliary variables c, b and *bdao* are introduced.

$$\forall c\ b\ bdao\ .\ \Gamma, \Theta \vdash \{|\ c = 'credit\ user\ \wedge\ b = balance\ ('gs\ user)\ \wedge$$
$$bdao = balance\ ('gs\ babyDao)\ |\}$$
$$withdraw$$
$$\{|\ 'credit\ user = 0\ \wedge\ balance\ ('gs\ user) = b + c\ \wedge$$
$$balance\ ('gs\ babyDao) = bdao - c\ |\}, \{\}$$

In the case of a so called friendly fallback function, the specification for *withdraw* holds. However, suppose an attacker writes a *fallback* function which besides increasing the attacker's balance and decreasing the balance of *babyDao*, contains code that checks whether the balance of *babyDao* will remain bigger than or equal to 0 after another possible withdraw, and if so, calls *withdraw*.

Suppose the attacker has some credit c and *bdao* is the total balance of *babyDao*. When the attacker calls the *withdraw* function, *call_value* transfers ether to the attacker, triggering its *fallback* function which may create another call to *withdraw*. This causes the attacker to receive the same amount of ether again and enter a recursive loop until all possible ether has been drained from *babyDao* without causing the function to fail, that is, the guard of the conditional statement in the *fallback* function never evaluates to false. The attacker's credit is only set to 0 after *babyDao*, and therefore, after all these recursive calls. The withdraw function ends up being called $\lfloor \frac{bdao}{c} \rfloor$ times and the attacker increases its value by $\lfloor \frac{bdao}{c} \rfloor \times c$.

In this case, the proof for the specification no longer holds but a proof for the attack can be established using the rules for multiple procedure recursive calls.

6   Conclusions

The main contribution from this work is the development of an imperative language and respective semantics system regarding a relevant subset of Solidity, based on a set of existent imperative languages in Isabelle/HOL, in particular the language proposed by Schirmer [23]. The main additions were the modelling of Solidity calls, both internal and external, Solidity exceptions, and reverting all state modifications. The relative completeness proof, based on the proof by Winskel [25], uses an auxiliary lemma that involves the concept of weakest precondition. After the addition of the *Dyncom, Require* and *Init* cases to the *wp* and *vc* computations, following the work by Frade and Pinto [7], we extend the proof of the lemma with the *Call, Handle, Revert, Dyncom, Require* and *Init* cases.

The main advantage of using a proof assistant is the richness with which properties about programs can be expressed as we saw in Sect. 5. From the *Ballot* example, it can be seen how invariants increase the complexity of a proof, but also how that complexity can be tackled using auxiliary properties. Also, the example of *Ethereum tokens* was analyzed and in most cases, upon a correct specification, the tactic *uint_arith* is able to find, or at least give a hint of, overflows and underflows. Finally, the possibility of recursion allows to model reentrancy vulnerabilities and fallback function attacks.

References

1. Amani, S., Bégel, M., Bortin, M., Staples, M.: Towards verifying Ethereum smart contract bytecode in Isabelle/HOL. In: CPP 2018, pp. 66–77. ACM (2018)
2. Bartoletti, M., Galletta, L., Murgia, M.: A minimal core calculus for solidity contracts. In: Pérez-Solà, C., Navarro-Arribas, G., Biryukov, A., Garcia-Alfaro, J. (eds.) DPM/CBT 2019. LNCS, vol. 11737, pp. 233–243. Springer, Cham (2019). https://doi.org/10.1007/978-3-030-31500-9_15
3. Bhargavan, K., et al.: Formal verification of smart contracts: short paper. In: PLAS 2016, pp. 91–96. ACM (2016)
4. Buterin, V.: Ethereum: a next-generation cryptocurrency and decentralized application platform
5. Cook, S.A.: Soundness and completeness of an axiom system for program verification. SIAM J. Comput. **7**, 70–90 (1978)
6. Dijkstra, E.W., Scholten, C.S.: Predicate Calculus and Program Semantics. Texts and Monographs in Computer Science. Springer, Heidelberg (1990). https://doi.org/10.1007/978-1-4612-3228-5
7. Frade, M.J., Pinto, J.S.: Verification conditions for source-level imperative programs. Comput. Sci. Rev. **5**(3), 252–277 (2011)
8. Grishchenko, I., Maffei, M., Schneidewind, C.: Foundations and tools for the static analysis of Ethereum smart contracts. In: Chockler, H., Weissenbacher, G. (eds.) CAV 2018. LNCS, vol. 10981, pp. 51–78. Springer, Cham (2018). https://doi.org/10.1007/978-3-319-96145-3_4

9. Grishchenko, I., Maffei, M., Schneidewind, C.: A semantic framework for the security analysis of Ethereum smart contracts. In: Bauer, L., Küsters, R. (eds.) POST 2018. LNCS, vol. 10804, pp. 243–269. Springer, Cham (2018). https://doi.org/10.1007/978-3-319-89722-6_10

10. Hildenbrandt, E., et al.: KEVM: a complete formal semantics of the Ethereum virtual machine. In: CSF 2018, pp. 204–217. IEEE Computer Society (2018)

11. Hirai, Y.: Defining the Ethereum virtual machine for interactive theorem provers. In: Brenner, M., et al. (eds.) FC 2017. LNCS, vol. 10323, pp. 520–535. Springer, Cham (2017). https://doi.org/10.1007/978-3-319-70278-0_33

12. Hoare, C.A.R.: An axiomatic basis for computer programming. Commun. ACM **12**(10), 576–580 (1969)

13. Hoare, C.A.R.: Procedures and parameters: an axiomatic approach. In: Engeler, E. (ed.) Symposium on Semantics of Algorithmic Languages. LNM, vol. 188, pp. 102–116. Springer, Heidelberg (1971). https://doi.org/10.1007/BFb0059696

14. Jiao, J., Kan, S., Lin, S., Sanán, D., Liu, Y., Sun, J.: Semantic understanding of smart contracts: executable operational semantics of solidity. In: SP 2020, pp. 1265–1282. IEEE Computer Society (2020)

15. Lincoln, P., Mitchell, J., Scedrov, A., Shankar, N.: Decision problems for propositional linear logic. Ann. Pure Appl. Logic **56**(1), 239–311 (1992)

16. Lincoln, P.D., Mitchell, J.C., Scedrov, A.: Linear logic proof games and optimization. Bull. Symbolic Logic **2**(3), 322–338 (1996)

17. Luu, L., Chu, D., Olickel, H., Saxena, P., Hobor, A.: Making smart contracts smarter. In: ACM CCS 2016, pp. 254–269. ACM (2016)

18. Mateus, P., Mitchell, J., Scedrov, A.: Composition of cryptographic protocols in a probabilistic polynomial-time process calculus. In: Amadio, R., Lugiez, D. (eds.) CONCUR 2003. LNCS, vol. 2761, pp. 327–349. Springer, Heidelberg (2003). https://doi.org/10.1007/978-3-540-45187-7_22

19. Mitchell, J.C., Ramanathan, A., Scedrov, A., Teague, V.: A probabilistic polynomial-time process calculus for the analysis of cryptographic protocols. Theor. Comput. Sci. **353**(1), 118–164 (2006)

20. Mythril. https://github.com/ConsenSys/mythril

21. Nakamoto, S.: Bitcoin: a peer-to-peer electronic cash system (2009)

22. Nikolic, I., Kolluri, A., Sergey, I., Saxena, P., Hobor, A.: Finding the greedy, prodigal, and suicidal contracts at scale. In: ACSAC 2018, pp. 653–663. ACM (2018)

23. Schirmer, N.: Verification of sequential imperative programs in Isabelle/HOL. Ph.D. thesis, Technical University Munich, Germany (2006)

24. Tsankov, P., Dan, A.M., Drachsler-Cohen, D., Gervais, A., Bünzli, F., Vechev, M.T.: Securify: practical security analysis of smart contracts. In: ACM CCS 2018, pp. 67–82. ACM (2018)

25. Winskel, G.: The Formal Semantics of Programming Languages: An Introduction. MIT Press, Cambridge (1993)

26. Wood, G.: Ethereum: a secure decentralised generalised transaction ledger. Ethereum Project Yellow Paper (2019)

27. Zakrzewski, J.: Towards verification of Ethereum smart contracts: a formalization of core of solidity. In: Piskac, R., Rümmer, P. (eds.) VSTTE 2018. LNCS, vol. 11294, pp. 229–247. Springer, Cham (2018). https://doi.org/10.1007/978-3-030-03592-1_13

Logic and Applications - LAP Meeting

Zvonimir Šikić[1], Silvia Ghilezan[2,3(✉)], Zoran Ognjanović[3],
and Thomas Studer[4]

[1] University of Zagreb, Zagreb, Croatia
[2] University of Novi Sad, Novi Sad, Serbia
gsilvia@uns.ac.rs
[3] Mathematical Institute SASA, Belgrade, Serbia
[4] University of Bern, Bern, Switzerland

The aim of this note is to bring your attention to Prof. Scedrov's role and impact in the foundation and shaping the practice of the scientific meeting Logic and Application as well as its regular organization.

The Logic and Application[1], a.k.a. LAP, is an annual meeting that brings together researchers from various fields of logic with applications in computer science, at large.

In 2012, Prof. Scedrov and the first three authors of this note agreed to meet in Dubrovnik during LiCS 2012, which was chaired by Prof. Scedrov, for a one-day workshop Sustavi dokazivanja (Proof systems). The main issue of the meeting was to discuss the possibilities to set-up a regular meeting that will bring together doctoral students from the region, early stage researchers and world-wide recognized experts in logic and its application. From the early 1970s, for almost twenty years, there was an ongoing seminar in mathematical logic between the research groups in Belgrade and Zagreb, which was alternating between the two venues. Prof. Scedrov, still a student at the time, was an active participant. Building on this good-practice and tradition, we have decided to set-up an annual meeting based at the Inter-University Center in Dubrovnik every September. Such kind of regular forum on research in logic was lacking in the region.

(a) Dubrovnik   (b) Inter University Center

From the very beginning, LAP was meant to have a very specific profile. It has been developed as a "slow meeting" focusing on detailed one hour expert

---

[1] The web page of LAP is http://imft.ftn.uns.ac.rs/math/cms/LAP.

© Springer Nature Switzerland AG 2020
V. Nigam et al. (Eds.): Scedrov Festschrift, LNCS 12300, pp. 98–100, 2020.
https://doi.org/10.1007/978-3-030-62077-6_8

talks, along with face-to-face interaction and exchange of ideas, as well as student presentations with joint discussions of peers and experienced researchers. It has been foreseen to have a schedule which enables, encourages and maximises the effectiveness in an environment which perfectly matches these requirements.

LAP is hosted by the Inter-University Center Dubrovnik[2], which is an independent international institution for advanced studies structured as a consortium of universities with a long standing tradition and mission to organise and promote contact and exchange through projects, study programmes, courses and conferences across a wide range of scientific concerns.

Topics of interest include, but are not restricted to: - Formal systems of classical and non-classical logic; - Category theory; - Proof theory; - Model theory; - Set theory; - Type theory; - Lambda calculus; - Process algebras and calculi; - Behavioural types; - Systems of reasoning in the presence of incomplete, imprecise and/or contradictory information; - Computational complexity; - Interactive theorem provers; - Security and Privacy.

(c) Prof. Scedrov's talk at LAP 2015

(d) LAP 2017

The role of Prof. Scedrov in the LAP meetings is multifold. Prof. Scedrov has initiated LAP; during nine years of the meeting, he always actively takes part in inviting lecturers and assembling the program; he gave talks at all eight LAP meetings and he has been thoroughly involved in advising doctoral students. Prof. Scedrov has the merits of making LAP a successful, lively and prosperous meeting.

(e) IUC reception, LAP 2018

(f) LAP co-directors, LAP 2019

---

[2] The web page of IUC is http://www.iuc.hr.

Since LAP 2013 the meeting has its current concept coordinated by five co-directors, Prof. Scedrov and the authors of this note. Since then there were eight editions and the organisation of the ninth edition, LAP 2020, is ongoing.

Welcome to LAP and meet with Prof. Scedrov!

Logic and Security

Logic and Economy

Formal Methods Analysis of the Secure Remote Password Protocol

Alan T. Sherman[1]([✉]), Erin Lanus[2], Moses Liskov[3], Edward Zieglar[4], Richard Chang[1], Enis Golaszewski[1], Ryan Wnuk-Fink[1], Cyrus J. Bonyadi[1], Mario Yaksetig[1], and Ian Blumenfeld[5]

[1] Cyber Defense Lab, University of Maryland, Baltimore County (UMBC), Baltimore, MD 21250, USA
sherman@umbc.edu
[2] Virginia Tech, Arlington, VA 22309, USA
lanus@vt.edu
[3] The MITRE Corporation, Burlington, MA 01720, USA
mliskov@mitre.org
[4] National Security Agency, Fort George G. Meade, MD 20755, USA
evziegl@nsa.gov
[5] Two Six Labs, Arlington, VA 22203, USA
ian.blumenfeld@twosixlabs.com

Abstract. We analyze the *Secure Remote Password (SRP)* protocol for structural weaknesses using the *Cryptographic Protocol Shapes Analyzer (CPSA)* in the first formal analysis of SRP (specifically, Version 3).

SRP is a widely deployed *Password Authenticated Key Exchange (PAKE)* protocol used in 1Password, iCloud Keychain, and other products. As with many PAKE protocols, two participants use knowledge of a pre-shared password to authenticate each other and establish a session key. SRP aims to resist dictionary attacks, not store plaintext-equivalent passwords on the server, avoid patent infringement, and avoid export controls by not using encryption. Formal analysis of SRP is challenging in part because existing tools provide no simple way to reason about its use of the mathematical expression $v + g^b \mod q$.

Modeling $v + g^b$ as encryption, we complete an exhaustive study of all possible execution sequences of SRP. Ignoring possible algebraic attacks, this analysis detects no major structural weakness, and in particular no leakage of any secrets. We do uncover one notable weakness of SRP, which follows from its design constraints. It is possible for a malicious server to fake an authentication session with a client, without the client's participation. This action might facilitate an escalation of privilege attack, if the client has higher privileges than does the server. We conceived of this attack before we used CPSA and confirmed it by generating corresponding execution shapes using CPSA.

Keywords: Cryptographic protocols · Cryptography · Cryptographic Protocol Shapes Analyzer (CPSA) · Cybersecurity · Formal methods · Password Authenticated Key Exchange (PAKE) protocols ·

© Springer Nature Switzerland AG 2020
V. Nigam et al. (Eds.): Scedrov Festschrift, LNCS 12300, pp. 103–126, 2020.
https://doi.org/10.1007/978-3-030-62077-6_9

Protocol analysis · Secure Remote Protocol (SRP) · UMBC Protocol
Analysis Lab (PAL)

1 Introduction

Cryptographic protocols underlie most everything that entities do in a networked
computing environment, yet, unfortunately, most protocols have never under-
gone any formal analysis. Until our work, this situation was true for the widely
deployed *Secure Remote Password (SRP)* protocol [28,51–53]. Given the complexity of protocols and limitations of the human mind, it is not feasible for
experts to find all possible structural flaws in a protocol; therefore, formal methods tools can play an important role in protocol analysis.

Protocols can fail for many reasons, including structural flaws, weak cryptography, unsatisfied hypotheses, improper configuration, inappropriate application, and implementation errors. We focus on structural weaknesses: fundamental
logic errors, which enable an adversary to defeat a protocol's security objective
or learn secret information.

We analyze SRP for structural weaknesses in the first formal analysis of
SRP (specifically, Version 3, known as *SRP-3*). Using the *Cryptographic Protocol
Shapes Analyzer (CPSA)* [36] tool in the *Dolev-Yao network intruder model* [21],
we model SRP-3 and examine all possible execution sequences of our model.
CPSA summarizes these executions with graphical "shapes," which we interpret.

SRP is a *Password Authenticated Key Exchange (PAKE)* protocol used in
1Password, iCloud Keychain, and other products. As with many PAKE protocols, two participants use knowledge of a pre-shared password to authenticate
each other and establish a session key. SRP aims to resist dictionary attacks,
not store plaintext-equivalent passwords on the server, avoid patent infringement, and avoid export controls by not using encryption.

Formal analysis of any protocol is challenging, and analysis of SRP is particularly difficult because of its use of the mathematical expression $v + g^b \mod q$.
This expression involves both modular exponentiation and modular addition,
exceeding the ability of automated protocol analysis tools to reason about modular arithmetic. Although SRP claims to have no encryption, ironically, we overcome this difficulty by modeling the expression as encryption, which, effectively
it is.

We created a new virtual protocol analysis lab at UMBC. Embodied as a
virtual machine running on the Docker utility,[1] this lab includes documentation,
educational modules for learning about protocol analysis, and three protocol
analysis tools: CPSA, Maude-NPA [24,25], and Tamarin Prover [22].

Contributions of our work include: (1) The first formal analysis of the SRP-3
protocol for structural weaknesses, which we carried out using the CPSA tool.
Ignoring possible algebraic attacks, this analysis detects no major structural
weakness, and in particular no leakage of any secrets. (2) The discovery of the
first attack on SRP, in which it is possible for a malicious server to fake an

---

[1] www.docker.com.

authentication session with the client, without the client's participation. This action might facilitate an escalation of privilege attack, if the client has higher privileges than does the server.

2   Background and Previous Work

We briefly review formal methods for analyzing cryptographic protocols, CPSA, PAKE protocols, and previous work on SRP.

2.1   Formal Methods for Analyzing Cryptographic Protocols

Several tools exist for formal analysis of cryptographic protocols, including CPSA [19,20,29], *Maude-NPA* [24,25], the Tamarin Prover [45], and ProVerif [8]. Created in 2005, CPSA outputs a set of "shapes" that describe all possible protocol executions, which can reveal undesirable execution states including ones caused by adversarial interference. Developed by Meadows [40] in 1992 as the NRL Protocol Analyzer, and rewritten into Maude language by Escobar et al. [23] in 2005, Maude-NPA works backwards from explicitly-defined attack states. The Tamarin Prover uses a multiset-rewriting model particularly well suited for analyzing stateful protocols. ProVerif is an automated cryptographic protocol verifier that operates on protocol specifications expressed in applied pi calculus, which specifications it translates into Horn clauses. We choose to use CPSA because we are more familiar with that tool, have easy access to experts, and like its intuitive graphical output.

A variety of additional tools exist to support formal reasoning, including for cryptography. For example, created in 2009, EasyCrypt[2] supports "reasoning about relational properties of probabilistic computations with adversarial code ... for the construction and verification of game-based cryptographic proofs." Cryptol [11] is a domain-specific language for cryptographic primitives. Cryptol allows for the symbolic simulation of algorithms, and thus the ability to prove properties of such by hooking into various constraint (SAT/SMT) solvers. Additionally, interactive theorem provers, such as Isabelle or Coq, have been used to analyze cryptographic functions and protocols [3,42]. These tools offer the potential to verify any property expressible in their underlying logics (higher-order logic or dependent type theory, respectively) but sacrifice automation.

The 1978 Needham-Schroeder [41] public-key authentication protocol dramatically illustrates the value of formal methods analysis and limitations of expert review. In 1995, using a protocol analysis tool, Lowe [38] identified a subtle structural flaw in Needham-Schroeder. This flaw had gone unnoticed for 17 years in part because Needham and Schroeder, and other security experts, had failed to consider the possibility that the intended recipient might be the adversary. Thus, for example, if Alice authenticates to Bob, then Bob could impersonate Alice to Charlie. CPSA easily finds this unexpected possible execution sequence, outputting a suspicious execution shape.

---

[2] https://www.easycrypt.info/trac/#no1.

Cryptographers sometimes present a *Universal Composability (UC)* proof of security [12], but such proofs as typically written are long and complex and can be difficult to verify. For example, Jarecki, Krawczyk, and Xu's [33] UC proof of the OPAQUE protocol is in a 61-page complex paper. There is, however, recent work on mechanically checking UC proofs (e.g., see Canetti, Stoughton, and Varia [13]), including Dolev-Yao versions of UC (e.g., see Böhl and Unruh [9] and Delaune, Kremer, and Pereria [16].) By contrast, to analyze SRP-3, CPSA requires only a relatively short and easy-to-verify input that formally defines the protocol in terms of its variables, the participant roles, and the messages sent and received.

2.2   Cryptographic Protocol Shapes Analyzer

The *Cryptographic Protocol Shapes Analyzer (CPSA)* [29,36,43] is an open-source tool for automated formal analysis of cryptographic protocols. The tool takes as input a model of a cryptographic protocol and a set of initial assumptions called the *point of view*, and attempts to calculate a set of minimal, essentially different executions of the protocol consistent with the assumptions. Such executions, called *shapes*, are relatively simple to view and understand. Executions in which something "bad" happens amount to illustrations of possible attacks against the protocol. Conversely, when some property holds in all shapes, it is a property guaranteed by the protocol.

CPSA is a tool based on *strand space theory* [20,26], which organizes events in a partially-ordered graph. In strand space theory, *events* are transmissions or receptions of messages, and sequences of events called *strands* capture the notion of the local viewpoint of a participant in a network. CPSA also has *state events*, which comprise initializing, observing, and transitioning between states. Protocols are defined as a set of legitimate participant roles, which serve as templates for strands consistent with the protocol requirements.

Bundles are the underlying execution model, in which every reception is explained directly by a previous transmission of that exact message. A bundle of a particular protocol is a bundle in which all the strands are either (1) generic adversary behavior such as parsing or constructing complex messages, or encrypting or decrypting with the proper keys, or (2) behavior of participants in the protocol consistent with the protocol roles.

CPSA reasons about bundles indirectly by analyzing *skeletons*, which are partially-ordered sets of strands that represent only regular behavior, along with origination assumptions that stand for assumptions about secrecy and/or freshness of particular values. For example, such assumptions might include that a key is never revealed or a nonce is freshly chosen and therefore assumed unique. Some skeletons represent, more or less, the exact set of regular behavior present in some bundle consistent with the secrecy and freshness assumptions; such skeletons are called *realized* skeletons. Realized skeletons are a simplified representation of actual protocol executions. Non-realized skeletons may represent partial descriptions of actual executions, or may represent a set of conditions inconsistent with any actual execution [36].

The CPSA tool creates visualizations of skeletons as graphs in which events are shown as circles in columns, where each column represents a strand. Within each strand, time progresses downward. *Arrows* between strands indicate necessary orderings (other than orderings within strands, or those that can be inferred transitively). That is, an arrow from event P to event Q denotes that, for Q to take place, it is necessary for P to take place first. A *solid arrow* represents a transmission of some message to a reception of exactly that message. A *dashed arrow* indicates that the adversary altered the message. The color of a circle indicates the type of event: *black circles* are transmissions; *blue circles* are receptions; and *grey circles* deal with state that is assumed to be not directly observable by the attacker. A *blue arrow* from state event P to state event Q denotes that Q's strand observes, or transitions from, the state associated with P; it can appear only between two state events (e.g., grey circles) of different strands. For example, Fig. 3 in Sect. 5.1 shows such a visualization.

2.3 PAKE Protocols

PAKE protocols evolved over time in response to new requirements and newly discovered vulnerabilities in authentication protocols [10]. Initially, authentication over a network was carried out simply with a username and password sent in the clear. Unlike terminals hardwired to a computer, networks provided new and easier ways for intruders to acquire authentication credentials. Passively monitoring a network often harvested credentials sufficient to gain remote access to systems. In the 1980's, *Kerberos* [47] attempted to mitigate this vulnerability by no longer transmitting passwords. Unfortunately, the structure of Kerberos messages and the use of passwords as keys created opportunities for password guessing and dictionary attacks against the passwords, without requiring the intruder to acquire the password file directly from the server. Weak, user-chosen passwords simplified such attacks.

In 1992, with their *Encrypted Key Exchange (EKE)* protocols, Bellovin and Merrit [6] evolved PAKE protocols to address the weaknesses in user-generated passwords as keys. In 1996, that work led Jablon [32] to develop the *Simple Password Exponential Key Exchange (SPEKE)*, which is deployed in the ISO/IEC 11770-4 and IEEE 1363.2 standards. As did Kerberos, to complicate dictionary attacks, SPEKE incorporated random *salt* values into its password computations. Attacks against the protocol in 2004 [54], 2005 [48], and 2014 [31], prompted modifications to the protocol. Although these and similar protocols aimed to protect against the use of weak passwords for authentication, none protected the passwords from attack on the server's password file. Access to the server's password file provided keys to authenticate as any user on the system.

Protection of the server's authentication file became a primary new requirement that Wu [51,52] aimed to address with the *Secure Remote Password (SRP)* protocol in 1998. Wu addressed this requirement by not storing the password, but instead a *verifier* consisting of a modular exponentiation of a generator raised to the power of a one-way hash function of the password. Improving on earlier

PAKE protocols, the way SRP incorporates a random salt into the key computation prevents the direct use of server-stored verifiers as keys. In 2002, weaknesses discovered against SRP-3 led Wu to propose a new version, *SRP-6* [53].

Unfortunately, for each password, SRP publicly reveals the corresponding salt, which facilitates pre-computation dictionary attacks on targeted passwords. Aware of this vulnerability, Wu nevertheless considered SRP a significant improvement over what had come before. Avoiding pre-computation attacks led to new approaches including the *OPAQUE* protocol [27,33,34].

2.4   Previous Work

SRP [49,50] is a widely used password-authenticated key-establishment protocol, which enables two communicants to establish a secret session key, provided the communicants already know a common password. SRP is faster than the authenticated Diffie-Hellman key exchange protocol, and it aims to avoid patent infringement and export control. In this protocol, an initiator Alice (typically a client) authenticates to a responder Bob (typically a server).

In this paper, we analyze the basic version of SRP called *SRP-3*. *SRP-6* mitigates a two-for-one attack and decreases communication times by allowing more flexible message orderings.

Against a passive adversary, SRP-3 seems to be as secure as the Diffie-Hellman problem [17,28,39]. It remains possible, however, that a passive adversary can acquire information from eavesdropping without solving the Diffie-Hellman problem. Against an active adversary, the security of SRP-3 remains unproven.

Wu [52] claims to prove a reduction from the Diffie-Hellman problem to breaking SRP-3 against a passive adversary, but his proof is incorrect: his reduction assumes the adversary knows the password, which a passive adversary would not know.[3] We are not aware of any other previous effort to analyze the SRP protocol.

Wilson et al. [7] survey authenticated Diffie-Helman key agreement protocols. Adrian et al. [1] analyze how such protocols can fail in practice. Schmidt et al. [45] present automated analysis of Diffie-Helman protocols.

As an example of formal analysis of a protocol using CPSA, we note: In 2009, Ramsdell et al. [43] analyzed the CAVES attestation protocol using CPSA, producing shapes that prove desirable authentication and confidentiality properties. The tool successfully analyzed the protocol despite the presence of hash functions and auxiliary long-term keys. As another example, which illustrates the utility of service roles, see Lanus and Zieglar [35]. Corin, Doumen, and Etalle [15] symbolically analyze offline guessing attacks.

---

[3] Wu incorrectly states the direction of his reduction, but his reduction actually proceeds in the correct direction.

3 The Secure Remote Password Protocol

Figure 1 summarizes how SRP-3 works, during which Alice and Bob establish a secret *session key* K, leveraging a *password* P known to Alice and Bob.

In SRP-3, all math is performed in some prime-order group $\mathbb{Z}_q$, where q is a large prime integer. Let g be a generator for this group. The protocol uses a hash function h. For brevity, for any $x \in \mathbb{Z}_q$, we shall write g^x to mean g^x mod q.

SRP-3 works in three phases: I. Registration. II. Key Establishment and III. Key Verification. The protocol establishes a new session key K known to Alice and Bob, which they can use, for example, as a symmetric encryption key.

Phase I works as follows: Before executing the protocol, Alice must register her password P with Bob. Bob stores the values (s, v) indexed by "Alice", where s is a random *salt*, $x = h(s, P)$ is the salted hash value of Alice's password, and $v = g^x$ is a non-sensitive *verifier* derived from P, which does not reveal x or P.

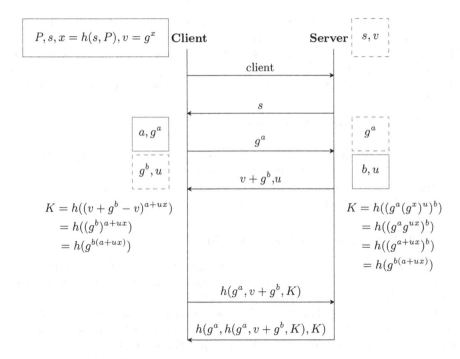

Fig. 1. Protocol diagram for SRP-3, which comprises three phases: Registration, Key Exchange, and Key Verification. During key exchange, the server transmits to the client the expression $v + g^b$ mod q, which we cannot directly model in CPSA. Variables on arrows inside the lattice diagram indicate message transmissions. Variables to the left or right of the lattice indicate terms known to the participants and the relative time within the protocol that they know them. Variables in solid boxes denote values chosen. Variables in dashed boxes denote values received.

Phase II works as follows:

1. Alice sends her identity "Alice" to Bob.
2. Bob receives Alice's identity and looks up Alice's salt s and stored verifier $v = g^x$, where $x = h(s, P)$. Bob sends Alice her salt s.
3. Alice receives s, calculates $x = h(s, P)$, and generates a random secret nonce a. Alice calculates and sends g^a to Bob.
4. Bob receives g^a and generates a random secret nonce b and a random scrambling parameter u. Bob calculates and sends $v + g^b$ to Alice, together with u.
5. Each party calculates the session key K as the hash of a common value, which each party computes differently. Alice calculates $K = h((v + g^b) - g^x)^{a+ux}$ and Bob calculates $K = h(g^a g^{ux})^b$.

Thus, in Phase II, Alice and Bob establish a common session key K. In Phase III, Alice and Bob verify that they have the same session key. Phase III works as follows:

1. Alice computes $M_1 = h(g^a, v + g^b, K)$ and sends M_1 to Bob. Bob verifies the received value by recomputing $M_1 = h(g^a, v + g^b, K)$.
2. Bob computes $M_2 = h(g^a, M_1, K)$ and sends it to Alice. Alice verifies the received value by recomputing $M_2 = h(g^a, M_1, K)$.
3. If and only if these two verifications succeed, the session key K is verified.

4   Modeling SRP-3 in CPSA

Using CPSA, we analyze SRP-3 in the Dolev-Yao network intruder model in two steps: in this section, we model SRP-3 in CPSA; in the next section, we interpret shapes produced by our model. Appendix A lists important snippets of our CPSA sourcecode.

4.1   Challenges to Modeling SRP-3 in CPSA

CPSA provides two algebras to express protocols: basic and Diffie-Hellman. The basic crypto algebra includes functions that support modeling of pairings, decomposing a pair into components, hashing, encrypting by symmetric and asymmetric keys, decrypting by keys, returning the "inverse of a key" (a key that can be used to decrypt), and returning a key associated with a name or pair of names. CPSA does not support arithmetic operations. The Diffie-Hellman algebra extends the basic crypto algebra by providing *sorts* (variable types) that represent exponents and bases, as well as functions for a standard generator g, a multiplicative identity for the group, exponentiation, and multiplication of exponents.

SRP-3 is challenging to model in CPSA because CPSA does not support any of the following computations: addition of bases when the server sends $v + g^b$, subtraction of bases when the client computes $(v + g^b) - v$, and addition of exponents (i.e., multiplication of bases) when the client computes the key. CPSA handles only multiplication of exponents, and cannot be easily modified to handle these additional algebraic operations, because CPSA makes use of general unifications in its class of messages, and a full decision procedure in the theory of rings is undecidable [14].

4.2   Our Model of SRP-3

We model SRP-3 by defining variables, messages, and associated roles. Critical modeling decisions are how to represent the problematic expression $v + g^b$, how to deal with multiplication of bases, and how to handle the initialization phase. Figure 2 shows the SRP-3 protocol diagram as we modeled SRP-3 in CPSA.

There are two legitimate protocol participants, which we model by the client and server roles (see Fig. 8). We organize each of these roles into two phases: initialization and main. The initialization phase establishes and shares the password, and it establishes the salt and verifier in the long-term memory of the server.

We model the problematic expression $v+g^b$ as $\{|g^b|\}_v$, which is the encryption of g^b using v as a symmetric key. Indeed, this modular addition resembles a Vernam Cipher. Thus, knowing g^b requires knowledge of v. Previous researchers have similarly modeled modular addition or exclusive-or as encryption (e.g., see Arapinis et al. [2] and Ryan and Schneider [44].)

The other problematic expressions occur in the calculation of the key. The key K is supposed to be equal to $(g^b)^{a+ux}$. Here, each party calculates this value by calculating g^{ab} and g^{bux} and multiplying them together. The client can calculate these values from g^b by raising g^b to the a power and to the ux power. The server calculates these values by raising g^a to the b power, and by raising $g^x = v$ to the bu power.

We emulate the multiplication of these base values by hashing them; since both parties can calculate the two factors, each can calculate the hash of the two factors. Thus, we represent the key K as $K = h(g^{ab}, g^{bux})$, where h stands for cryptographic hashing.

Finally, we explain how we model the initialization phase, and in particular, how the client communicates their salt and verifier to the server. In the beginning of the client and server roles, one could exchange the salt and verifier as a message. This strategy, however, would prevent CPSA from exploring scenarios in which the same client or server conducts multiple executions of the protocol using the same password information exchanged during initialization. Instead, we model the initialization phase using *service roles*, which provide a function or service to one or more participant roles. Our service roles generate values, store them in state, and exchange the values across a secure channel. These values persist in state that can be accessed only by instances of the appropriate main-phase roles.

Specifically, the client-init service role initializes a state record with the value { "client state", s, x, client, server} (see Fig. 8). The "client state" string literal serves the function of a label, enabling us to write client roles to observe state that begins with that string. We store the salt and password hash because each client role directly uses these values. The names of the client and server help to link the state to the correct client-server pair.

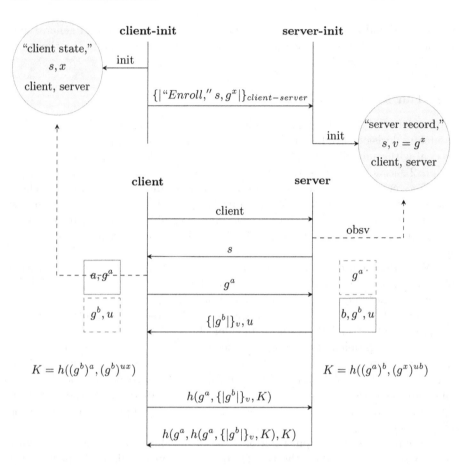

Fig. 2. Protocol diagram for SRP-3, as we modeled it in CPSA. We introduce two service roles, client-init and server-init, that handle the setup phase by instantiating values for s, x, and $v = g^x$, and by making these values available to the legitimate client and server. We model the computation $v + g^b$ as an encryption of g^b under key v. The red circle indicates the variables stored inside the state. Solid lines pointing to the circles denote initializing state values, and dashed lines indicate observing state. (Color figure online)

After initializing its state, the client-init role sends a string literal "Enroll", together with the salt and verifier. The client-init role encrypts this message using a long-term key known by the particular client and server. The server-init role receives this message and initializes the server's state by storing a string literal "server record", the salt and verifier, and the names of the client and server.

To prevent CPSA from instantiating an unlimited number of server-init and client-init roles, we add a rule that disregards any executions in which there is more than one instance of the server-init role for a specific client-server pair (see Fig. 9).

The model above is sufficient to verify most of the security properties of SRP, but cannot verify the property that compromise of the server's authentication database cannot be used directly to gain immediate access to the server. The reason is that if SRP meets its security goals, the verifier v is not leaked to the adversary by the protocol. Therefore, to test whether or not access to v allows the adversary to impersonate a client to the server, we need to use a model in which the server-init role is modified to transmit the verifier it receives for a client. This model provides the adversary with access to v that they cannot obtain from SRP. For this property, it is sufficient to test only the server's point of view. Compromise of a server's authentication database would allow anyone to impersonate a server to the client and is not a property that SRP was designed to prevent.

5   Interpreting Shapes from the SRP-3 Model

We generate and interpret shapes showing executions of our model of SRP-3 under various assumptions from the perspectives of various roles. Specifically, we define skeletons that provide the perspectives of an honest client and an honest server, respectively (see Figs. 10 and 11). We also define *listeners* to detect possible leaked values of the password hash x or verifier v (see Figs. 12 and 13). Finally, we investigate if an adversary directly using a compromised verifier could authenticate as a client (see Fig. 6). CPSA completed its search, generating all possible shapes for each point of view (see [37] for an explanation).

Figures 3, 4, 5 and 6 display selected shapes that highlight our main findings. These shapes show that, when the client and server are honest, there is no attack against our model of SRP-3: the only way the protocol completes is between a client and a server. Similarly, CPSA found no leakage of x or v. CPSA also found that an adversary directly using a compromised verifier cannot authenticate as a client without access to internal values of the server.

Our public GitHub repository [46] includes interactive web-based visualizations of our CPSA shapes and skeletons, which provide more detailed information than do the static images in this paper.

5.1   Client Point of View

Figures 3 and 4 show the two shapes generated from the perspective of an honest client. The first shape is what we had expected. One added client-init strand provides state needed for the client to access password information, and one added server-init strand provides password information to the server strand. The solid lines in the shape prove that the messages must come from the expected parties, and the shape closely reflects the protocol diagram for our model.

The second shape explores the possibility that the adversary could replay the client's initial message to the server resulting in the server beginning two protocol runs with the client. We are able to verify that it is the same server by observing that the server variables in both strands are instantiated with the same value.

Only one of the server strands is able to complete, because the messages between the two runs of the protocol cannot be confused. The shape indicates that there is not any way for the adversary to take advantage of initiating multiple runs of the protocol with the server.

srp3 22 (realized)

Fig. 3. Shape showing an execution of SRP-3 from the client's perspective. The client-init service role begins the execution. The blue arrow from the client-init strand to the client strand denotes that the client observes the initial state from client-init. Similarly, the blue arrow from the server-init strand to the server strand denotes that the server observes state from server-init. Horizontal black arrows between the client and server represent successful message transmissions and receptions between these two protocol participants. This graphical output from CPSA reveals expected behavior. (Color figure online)

5.2   Server Point of View

Figure 5 shows the first of two shapes generated from the perspective of an honest server. As happens for the client, two shapes result. The first shape is similar to the protocol diagram for our model and is what we had expected. A client is needed to complete the protocol, as are the service roles server-init and client-init. The second shape indicates a replay of the client's initial message resulting in two server strands with the same server as indicated in the strands' instantiated variables. As with the additional shape in the client's view, only one of the server's strands is able to complete, indicating that there is no attack against the protocol from the server point of view.

srp3 41 (realized)

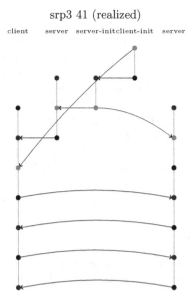

Fig. 4. Shape showing an execution of SRP-3 from the client's perspective, with an additional run of the server. This graphical output from CPSA reveals two server roles accessing the same state, causing them to behave like two instances of the same server. The client can begin the protocol with one instance of the server, then complete it with the other. This intriguing shape does not suggest any harmful attack but is an unavoidable consequence of CPSA exploring two server strands.

srp3 5119 (realized)

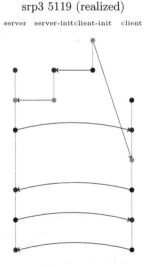

Fig. 5. Shape showing an execution of SRP-3 from the server's perspective. This figure is similar to Fig. 3, except CPSA is now trying to explain the server events. CPSA is able to explain the server events only by involving client-init, server-init, and client roles, thus revealing expected behavior.

5.3   Privacy Properties

It is important that the password hash $x = h(s, P)$ and the verifier $v = g^x$ remain secret. To determine whether a network adversary can observe either of these values in our model of SRP-3, we define two input skeletons to test these privacy properties, one for x and one for v (see Figs. 12 and 13). Because the client knows x, we add the listener for x to the client point of view. Similarly, because the server knows v, we add the listener for v to the server point of view. Listeners in CPSA represent a test that a value can be found by the adversary.

For each of these skeletons, we ran CPSA. In each case, CPSA returned an empty tree, meaning that there is no way to realize the skeleton as a shape, which means that no such attack is possible in our model. In each case, CPSA ran to completion, indicating that it explored all possible shapes for the model.

5.4   Leaked Verifiers

CPSA analysis of listeners for v confirms that the SRP protocol does not leak the verifier v. Therefore, to analyze the protocol when the adversary has access to v, we modified server-init to leak the verifier to the adversary. In the presence of this variant of the server-init role, CPSA discovered two main shapes: one is the ordinary server point of view (Fig. 5); the other shows that the adversary is able to impersonate a client if the verifier has indeed leaked (Fig. 6).

srp 7022 (realized)

server   client-initserver-init

Fig. 6. Shape showing an execution of SRP-3 from the server's perspective, when the verifier is leaked to the adversary and $u = b$. It is suspicious that CPSA can explain all server events without invoking the client. In the last event of the server-init strand, the server-init leaks the verifier to the adversary. The dashed arrow indicates that the adversary is able to use the leaked verifier, together with their knowledge of b (since u is publicly known), to satisfy the server strand's final event, and complete the protocol. This shape indicates an attack where the adversary impersonates the client to the server, when the adversary learns the verifier and b.

The situation is more subtle. The adversary is able to impersonate the client only if they know both v and b, as an adversary might learn if the adversary comprised the server. Initially, in our model of SRP-3, we did not require that b and u be distinct, only that they be uniquely generated. CPSA found the impersonation attack in part because CPSA deduced that the adversary could learn b if $b = u$, since SRP-3 reveals u. Subsequently, when we added an additional assumption that $b \neq u$, CPSA discovered only the expected shapes. This fact validates the assertion that SRP is secure from an adversary directly using the verifier to authenticate as a client without access to internal values of the server.

6   A Malicious Server Attack Against SRP

Our analysis in Sect. 5 assumes that legitimate participants of SRP-3 are honest, meaning they will execute the protocol faithfully. In this section, we explore an attack on SRP-3 in which the server is compromised. For example, an adversary might corrupt the server to run a malicious process. In this attack, the malicious server authenticates to itself, pretending to be a particular client, without the client's involvement. A possible goal of this attack might be for the malicious server to escalate its privileges to those of the client, which might be higher than those of the server. For example, a company might have a high-power, low-trust offline computing server used by individuals with sensitive access elsewhere in the network.

To analyze this attack, we define a malicious server role, which we call *malserver* (see Fig. 14). We provide to malserver only the information that an honest server would have access to by observing the state initialized by a server-init role. Consequently, malserver must compute the key using the same method as carried out by an honest server. Malserver also acts like a client, initiating the protocol and sending messages consistent with those from the client role. Figure 14 also defines an associated skeleton, which enables CPSA to compute a strand of the malserver role.

Figure 7 shows the first of two shapes produced by CPSA from the malserver skeleton. As for honest participants, CPSA also produced a second shape that shows the protocol can be started and completed with two different honest server roles on the same machine. Figure 7 shows the malserver initiating the protocol by sending the client's name and proceeding to interact with the server as though it were the client, all the way through to the key verification messages. For executions with a legitimate client, CPSA adds client-init and server-init strands, as a result of the setup phase in which a client sends name, salt, and verifier to the server. Here, however, there is no client strand. The server sends the final black node on its strand only after the server verifies the hash provided by the malserver strand, indicating that the server believes it is communicating with the specified client.

The attack is possible because the malserver role is operating on the server it is attacking (the server and malserver variables are equal) and has access to the server's internal values, as we discuss in the analysis of the leaked verifiers.

srp3 23 (realized)

Fig. 7. Shape showing an execution of SRP-3 from the perspective of a malicious server impersonating the client. It is notable that CPSA can explain all events of the malicious server strand, simply by the malicious server knowing the state of the honest server (without the malicious server knowing the client's password). This graphical output from CPSA reveals that a malicious server can impersonate the client to itself (the server), thereby potentially inheriting the client's higher privileges.

Even though this attack is not a part of the Dolev-Yao model that CPSA uses, by creating a special malserver role outside of the normal protocol roles, we were able to coax CPSA to explore the attack. This approach is similar to work by Basin and Cremers [4,5].

7   Discussion

We briefly discuss two limitations of our work: one arising from our modeling of the problematic expression $v + g^b$ as encryption, the other arising from our choice of CPSA's point of view (see Sect. 2.2). We also comment briefly on our experiences using CPSA.

Modeling the problematic expression as encryption enabled CPSA to carry out its work. A consequence of this crucial decision, however, is that we analyzed a slight variation of SRP-3 that might be stronger than SRP-3. By abstracting these algebraic operations as strong encryption, our analysis cannot find possible "algebraic attacks" that might take advantage of detailed algebraic relationships. We are not aware of any such attacks on SRP-3 and do not suspect that they exist, but we cannot exclude their possible existence. The consequences of this crucial modeling decision are similar to those from the common practice of modeling a particular encryption function as a strong encryption function, which excludes the possibility of finding attacks that exploit possible weaknesses in the particular encryption function.

CPSA exhaustively explores possible executions of a protocol from a specified point of view and set of assumptions. Such analysis holds only when those assumptions are satisfied for that point of view. For example, initially, CPSA did not find the malicious server attack described in Sect. 6. CPSA did not find this attack because the adversary requires access to variables v and b, that are not available through the messages exchanged and the assumptions of the model. We were able to show that SRP-3 does not leak those values. Similarly, initially, CPSA could not verify SRP-3's property that access to the state variable v by the adversary would not allow the adversary to impersonate a client directly. To verify that property would require a model that made v available to the adversary.

Subsequently, we explored two models to investigate possible impersonation attacks. One model gave the adversary v; the other model gave the adversary v and b. With these models, CPSA showed that the adversary can impersonate the client if they know v and b, but not if they know only v (see Sect. 5).

Different assumptions and points of view can influence analyses. All formal methods tools explore properties only within a specified scope and do not find attacks outside that scope. Although CPSA did not initially discover the malicious server attack, we were able to enlarge CPSA's scope of search to find it. It is possible, however, that there might be additional attacks outside our scope of search.

During our analysis of SRP-3, the graphical outputs of CPSA helped us gain insights into the properties of the protocol. Nevertheless, using CPSA effectively was challenging. It required learning a new complex language, gaining experience interpreting shapes, devising techniques to model algebraic expressions that cannot be expressed directly in CPSA, and exploring ways to expand CPSA's point of view. Embodied as a virtual machine, our virtual protocol lab avoids the need for users to carry out complex installation procedures for each tool.

We found the following existing techniques useful. (1) Service roles (e.g., client-init, server-init) permitted us to share state between protocol participants (e.g., server, client) and, more generally, to model aspects of protocols that do not directly involve communications among participants. (2) We modeled certain algebraic expressions as basic cryptographic operations such as encryption or hashing. (3) Defining additional protocol participants (e.g., listeners, malicious server) enabled us to explore additional properties of SRP-3 and to expand the capabilities of the Dolev-Yao adversary. We are sharing theses and other lessons in our lab's educational materials.

8   Conclusion

Using CPSA, we formally analyzed the SRP-3 protocol in the Dolev-Yao network intruder model and found it free of major structural weakness. We did find a weakness that a malicious server can fake an authentication session with a client without the client's participation, which might lead to an escalation of privilege attack.

Limitations of our analysis stem in part from our cryptographic modeling. CPSA will not find attacks that exploit weak cryptography, and our use of CPSA will not find any algebraic attacks. As all tool users must, we trust the correctness of CPSA and its execution. Our results do not speak to a variety of other potential issues, including possible implementation and configuration errors when using SRP-3, inappropriate applications of it, and side-channel attacks.

Open problems include formal analysis of other PAKE protocols [30], including the recent OPAQUE protocol [27,33,34], which, unlike SRP, tries to resist precomputation attacks by not revealing the salt values used by the server. OPAQUE is the most promising new protocol possibly to replace SRP. Because quantum computers can compute discrete logarithms in polynomial time, it would be useful to study and develop post-quantum PAKE protocols [18] that can resist quantum attack.

We hope that our work, as facilitated by the virtual protocol analysis lab created at UMBC, will help raise the expectation of due diligence to include formal analysis when designing, standardizing, adopting, and evaluating cryptographic protocols.

Acknowledgments. We appreciate the helpful comments from Akshita Gorti and the reviewers. Thanks also to John Ramsdell (MITRE) and other participants at the Protocol eXchange for fruitful interactions. This research was supported in part by the U.S. Department of Defense under CySP Capacity grants H98230-17-1-0387, H98230-18-1-0321, and H98230-19-1-0308. Sherman, Golaszewski, Wnuk-Fink, Bonyadi, and the UMBC Cyber Defense Lab were supported also in part by the National Science Foundation under SFS grants DGE-1241576, 1753681, and 1819521.

To appear in Festschrift in Honour of Professor Andre Scedrov, Vivek Nigam, Editor, LNCS, Springer (June 11, 2020).

A   CPSA Sourcecode

We list critical snippets of CPSA sourcecode that we used to model SRP-3 and carry out our analysis. A complete electronic version is available from our public GitHub repository [46].

```
(defprotocol srp3 diffie-hellman

  (defrole client-init
    (vars (s text) (x rndx) (client server name))
    (trace
     (init (cat "Client state" s x client server))
     (send (enc "Enroll" s (exp (gen) x) client (ltk client server))))
    (uniq-gen s x))

  (defrole server-init
    (vars (s text) (v mesg) (client server name))
    (trace
     (recv (enc "Enroll" s v client (ltk client server)))
     (init (cat "Server record" s v client server))))

  (defrole client
    (vars (client server name) (a rndx) (b u x expt) (s text))
    (trace
     (send client)
     (recv s)
     (obsv (cat "Client state" s x client server))
     (send (exp (gen) a))
     (recv (cat (enc (exp (gen) b) (exp (gen) x)) u))
     (send (hash (exp (gen) a)
               (enc (exp (gen) b) (exp (gen) x)) u
               (hash (exp (gen) (mul b a)) (exp (gen) (mul b u x)))))
     (recv (hash (exp (gen) a)
               (hash (exp (gen) a)
                     (enc (exp (gen) b) (exp (gen) x)) u
                     (hash (exp (gen) (mul b a)) (exp (gen) (mul b u x))))
               (hash (exp (gen) (mul b a)) (exp (gen) (mul b u x))))))
    (uniq-gen a))

  (defrole server
    (vars (client server name) (a expt) (b u rndx) (s text) (v base))
    (trace
     (recv client) ; Server receives Client's name
     (obsv (cat "Server record" s v client server))
     (send s)
     (recv (exp (gen) a))
     (send (cat (enc (exp (gen) b) v) u))
     (recv (hash (exp (gen) a)
               (enc (exp (gen) b) v) u
               (hash (exp (gen) (mul a b)) (exp v (mul u b)))))
     (send (hash (exp (gen) a)
               (hash (exp (gen) a)
                     (enc (exp (gen) b) v) u
                     (hash (exp (gen) (mul a b)) (exp v (mul u b))))
               (hash (exp (gen) (mul a b)) (exp v (mul u b))))))
    (uniq-gen u b))
)
```

Fig. 8. Modeling of SRP-3 in CPSA. We define four roles: client-init, server-init, client, and server. The client-init and server-init roles are service roles that initialize common values between the client and server roles.

```
(defrule at-most-one-server-init-per-client
  (forall ((z0 z1 strd) (client server name))
          (implies
            (and (p "server-init" z0 1)
                 (p "server-init" z1 1)
                 (p "server-init" "client" z0 client)
                 (p "server-init" "client" z1 client)
                 (p "server-init" "server" z0 server)
                 (p "server-init" "server" z1 server))
          (= z0 z1))
)
```

Fig. 9. Rule added to SRP-3 to prevent CPSA from instantiating an unlimited number of server-init roles, which would prevent CPSA from terminating.

```
(defskeleton srp3
  (vars (client server name))
  (defstrand client 7 (server server) (client client))
  (non-orig (ltk client server)))
```

Fig. 10. Client skeleton of SRP-3, which provides CPSA a starting point for analyzing SRP-3 from the client's perspective.

```
(defskeleton srp3
  (vars (client server name))
  (defstrand server 7 (server server) (client client))
  (non-orig (ltk client server)))
```

Fig. 11. Server skeleton of SRP-3, which provides CPSA a starting point for analyzing SRP-3 from the server's perspective.

```
(defskeleton srp3
  (vars (client server name))
  (defstrand client 7 (server server) (client client))
  (deflistener x)
  (non-orig (ltk client server)))
```

Fig. 12. Client skeleton of SRP-3 with listener for the value x, which provides CPSA a starting point for analyzing SRP-3 from the client's perspective. The listener role helps CPSA determine whether an execution of SRP-3 can leak the value x.

```
(defskeleton srp3
  (vars (client server name))
  (defstrand server 7 (server server) (client client))
  (deflistener v)
  (non-orig (ltk client server)))
```

Fig. 13. Server skeleton of SRP-3 with listener for the value v, which provides CPSA a starting point for analyzing SRP-3 from the server's perspective. The listener role helps CPSA determine whether an execution of SRP-3 can leak the value v.

```
(defrole malserver
    (vars (client server name) (a rndx) (b u expt) (s text) (v base))
    (trace
     (send client)
     (recv s)
     (obsv (cat "Server record" s v client server))
     (send (exp (gen) a))
     (recv (cat (enc (exp (gen) b) v) u))
     (send (hash (exp (gen) a)
                 (enc (exp (gen) b) v) u
                 (hash (exp (gen) (mul a b)) (exp v (mul u b)))))
     (recv (hash (exp (gen) a)
                 (hash (exp (gen) a)
                       (enc (exp (gen) b) v) u
                       (hash (exp (gen) (mul a b)) (exp v (mul u b))))
                 (hash (exp (gen) (mul a b)) (exp v (mul u b))))))
    (uniq-gen a)
)

(defskeleton srp3
    (vars (client server name))
    (defstrand malserver 7 (server server) (client client))
    (non-orig (ltk client server)))
```

Fig. 14. Modeling a malicious server in CPSA. We define the malserver role to behave like a client while having access to the legitimate server's initialized variables. The associated skeleton provides CPSA a starting point for analyzing the malicious server attack from the perspective of the malicious server.

References

1. Adrian, D., et al.: Imperfect forward secrecy: how Diffie-Hellman fails in practice. In: Proceedings of the 22nd ACM SIGSAC Conference on Computer and Communications Security, CCS 2015, pp. 5–17. ACM, New York (2015). https://doi.org/10.1145/2810103.2813707

2. Arapinis, M., et al.: New privacy issues in mobile telephony: fix and verification. In: Proceedings of the 2012 ACM Conference on Computer and Communications Security, CCS 2012, pp. 205–216. Association for Computing Machinery, New York (2012). https://doi.org/10.1145/2382196.2382221

3. Bartzia, E.-I., Strub, P.-Y.: A formal library for elliptic curves in the Coq proof assistant. In: Klein, G., Gamboa, R. (eds.) ITP 2014. LNCS, vol. 8558, pp. 77–92. Springer, Cham (2014). https://doi.org/10.1007/978-3-319-08970-6_6

4. Basin, D., Cremers, C.: Modeling and analyzing security in the presence of compromising adversaries. In: Gritzalis, D., Preneel, B., Theoharidou, M. (eds.) ESORICS 2010. LNCS, vol. 6345, pp. 340–356. Springer, Heidelberg (2010). https://doi.org/10.1007/978-3-642-15497-3_21

5. Basin, D., Cremers, C.: Know your enemy: compromising adversaries in protocol analysis. ACM Trans. Inf. Syst. Secur. **17**(2) (2014). https://doi.org/10.1145/2658996

6. Bellovin, S.M., Merritt, M.: Encrypted key exchange: password-based protocols secure against dictionary attacks. In: IEEE Symposium on Research in Security and Privacy, pp. 72–84, May 1992
7. Blake-Wilson, S., Menezes, A.: Authenticated Diffie-Hellman key agreement protocols. In: Proceedings of the Selected Areas in Cryptography, SAC 1998, pp. 339–361. Springer, Heidelberg (1999). http://dl.acm.org/citation.cfm?id=646554.694440
8. Blanchet, B., Smyth, B., Cheval, V.: Proverif 1.90: automatic cryptographic protocol verifier, user manual and tutorial (2015). http://prosecco.gforge.inria.fr/personal/bblanche/proverif/manual.pdf
9. Böhl, F., Unruh, D.: Symbolic universal composability. J. Comput. Secur. **24**(1), 1–38 (2016)
10. Boneh, D., Shoup, V.: A graduate course in applied cryptography version 0.5, January 2020. https://crypto.stanford.edu/~dabo/cryptobook/BonehShoup_0_5.pdf
11. Browning, S.: Cryptol, a DSL for cryptographic algorithms. In: ACM SIGPLAN Commercial Users of Functional Programming, p. 1. ACM (2010)
12. Canetti, R.: Universally composable security: a new paradigm for cryptographic protocols. In: Proceedings of the 42nd IEEE Symposium on Foundations of Computer Science, FOCS 2001, p. 136. IEEE Computer Society, USA (2001)
13. Canetti, R., Stoughton, A., Varia, M.: EasyUC: using EasyCrypt to mechanize proofs of universally composable security. In: 2019 IEEE 32nd Computer Security Foundations Symposium (CSF), pp. 167–183 (2019)
14. Church, A.: An unsolvable problem of elementary number theory. Am. J. Math. **58**(2), 345–363 (1936)
15. Corin, R., Doumen, J., Etalle, S.: Analysing password protocol security against off-line dictionary attacks. Electron. Notes Theoret. Comput. Sci. **121**, 47–63 (2005)
16. Delaune, S., Kremer, S., Pereira, O.: Simulation based security in the applied pi calculus. In: Kannan, R., Kumar, K.N. (eds.) IARCS Annual Conference on Foundations of Software Technology and Theoretical Computer Science. Leibniz International Proceedings in Informatics (LIPIcs), vol. 4, pp. 169–180. Schloss Dagstuhl-Leibniz-Zentrum fuer Informatik, Dagstuhl, Germany (2009). http://drops.dagstuhl.de/opus/volltexte/2009/2316
17. Diffie, W., Hellman, M.: New directions in cryptography. IEEE Trans. Inf. Theor. **22**(6), 644–654 (2006). https://doi.org/10.1023/A:1008302122286
18. Ding, J., Alsayigh, S., Lancrenon, J., RV, S., Snook, M.: Provably secure password authenticated key exchange based on RLWE for the post-quantum world. In: Handschuh, H. (ed.) CT-RSA 2017. LNCS, vol. 10159, pp. 183–204. Springer, Cham (2017). https://doi.org/10.1007/978-3-319-52153-4_11
19. Doghmi, S., Guttman, J., Thayer, F.J.: Skeletons and the shapes of bundles. In: Proceedings of the 7th International Workshop on Issues in the Theory of Security, pp. 24–25 (2006)
20. Doghmi, S.F., Guttman, J.D., Thayer, F.J.: Searching for shapes in cryptographic protocols. In: Grumberg, O., Huth, M. (eds.) TACAS 2007. LNCS, vol. 4424, pp. 523–537. Springer, Heidelberg (2007). https://doi.org/10.1007/978-3-540-71209-1_41
21. Dolev, D., Yao, A.C.: On the security of public key protocols. In: Proceedings of the 22nd Annual Symposium on Foundations of Computer Science, SFCS 1981, pp. 350–357. IEEE Computer Society, Washington, DC (1981). https://doi.org/10.1109/SFCS.1981.32

22. Dreier, J., Duménil, C., Kremer, S., Sasse, R.: Beyond subterm-convergent equational theories in automated verification of stateful protocols. In: Maffei, M., Ryan, M. (eds.) POST 2017. LNCS, vol. 10204, pp. 117–140. Springer, Heidelberg (2017). https://doi.org/10.1007/978-3-662-54455-6_6. https://hal.inria.fr/hal-01430490/document

23. Escobar, S., Meadows, C., Meseguer, J.: A rewriting-based inference system for the NRL protocol analyzer and its meta-logical properties. Theoret. Comput. Sci. **367**(1–2), 162–202 (2006)

24. Escobar, S., Meadows, C., Meseguer, J.: Maude-NPA: cryptographic protocol analysis modulo equational properties. In: Aldini, A., Barthe, G., Gorrieri, R. (eds.) FOSAD 2007-2009. LNCS, vol. 5705, pp. 1–50. Springer, Heidelberg (2009). https://doi.org/10.1007/978-3-642-03829-7_1

25. Escobar, S., Meadows, C., Meseguer, J.: Maude-NPA, Version 3.0, April 2017

26. Fabrega, F.J.T., Herzog, J.C., Guttman, J.D.: Strand spaces: why is a security protocol correct? In: Proceedings of the 1998 IEEE Symposium on Security and Privacy (Cat. No. 98CB36186), pp. 160–171, May 1998. https://doi.org/10.1109/SECPRI.1998.674832

27. Green, M.: Let's talk about PAKE, October 2018. https://blog.cryptographyengineering.com/2018/10/19/lets-talk-about-pake/

28. Green, M.: Should you use SRP? October 2018. https://blog.cryptographyengineering.com/should-you-use-srp/

29. Guttman, J.D., Liskov, M.D., Ramsdell, J.D., Rowe, P.D.: The Cryptographic Protocol Shapes Analyzer (CPSA). https://github.com/mitre/cpsa

30. Haase, B., Labrique, B.: AuCPace: Efficient verifier-based PAKE protocol tailored for the IIoT. IACR Trans. Cryptogr. Hardw. Embed. Syst. **2019**, 1–48 (2018)

31. Hao, F., Shahandashti, S.F.: The SPEKE protocol revisited. In: Chen, L., Mitchell, C. (eds.) SSR 2014. LNCS, vol. 8893, pp. 26–38. Springer, Cham (2014). https://doi.org/10.1007/978-3-319-14054-4_2

32. Jablon, D.P.: Strong password-only authenticated key exchange. ACM Comput. Commun. Rev. **26**(5), 5–26 (1996)

33. Jarecki, S., Krawczyk, H., Xu, J.: OPAQUE: An asymmetric PAKE protocol secure against pre-computation attacks. Cryptology ePrint Archive, Report 2018/163 (2018). https://eprint.iacr.org/

34. Jarecki, S., Krawczyk, H., Xu, J.: OPAQUE: an asymmetric PAKE protocol secure against pre-computation attacks. In: Nielsen, J.B., Rijmen, V. (eds.) EUROCRYPT 2018. LNCS, vol. 10822, pp. 456–486. Springer, Cham (2018). https://doi.org/10.1007/978-3-319-78372-7_15

35. Lanus, E., Zieglar, E.: Analysis of a forced-latency defense against man-in-the-middle attacks. J. Inf. Warfare **16**(2), 66–78 (2017). https://www.jstor.org/stable/26502758

36. Liskov, M.D., Ramsdell, J.D., Guttman, J.D., Rowe, P.D.: The Cryptographic Protocol Shapes Analyzer: A Manual. The MITRE Corporation (2016)

37. Liskov, M.D., Rowe, P.D., Thayer, F.J.: Completeness of CPSA. Technical report MTR110479, The MITRE Corporation (2011)

38. Lowe, G.: An attack on the Needham-Schroeder public-key authentication protocol. Inf. Process. Lett. **56**(3), 131–133 (1995). http://www.sciencedirect.com/science/article/pii/0020019095001442

39. Maurer, U.M., Wolf, S.: The Diffie-Hellman protocol. Des. Codes Cryptography **19**(2–3), 147–171 (2000). https://doi.org/10.1023/A:1008302122286

40. Meadows, C.: NRL protocol analyzer. J. Comput. Secur. **1**(1) (1992)

41. Needham, R.M., Schroeder, M.D.: Using encryption for authentication in large networks of computers. Commun. ACM **21**(12), 993–999 (1978). https://doi.org/10.1145/359657.359659

42. Paulson, L.C.: Relations between secrets: two formal analyses of the Yahalom protocol. J. Comput. Secur. **9**(3), 197–216 (2001)

43. Ramsdell, J.D., Guttman, J.D., Millen, J.K., O'Hanlon, B.: An analysis of the CAVES attestation protocol using CPSA. arXiv preprint arXiv:1207.0418 (2012)

44. Ryan, P.Y.A., Schneider, S.A.: An attack on a recursive authentication protocol. A cautionary tale. Inf. Process. Lett. **65**(1), 7–10 (1998). https://doi.org/10.1016/S0020-0190(97)00180-4

45. Schmidt, B., Meier, S., Cremers, C., Basin, D.: Automated analysis of Diffie-Hellman protocols and advanced security properties. In: 2012 IEEE 25th Computer Security Foundations Symposium, pp. 78–94, June 2012

46. Sherman, A.T., et al.: PAL GitHub repository, June 2020. https://github.com/egolaszewski/UMBC-Protocol-Analysis-Lab

47. Steiner, J.G., Neuman, B.C., Schiller, J.I.: Kerberos: an authentication service for open network systems. In: Proceedings Winter USENIX Conference, pp. 191–202 (1988)

48. Tang, Q., Mitchell, C.J.: On the security of some password-based key agreement schemes. In: Hao, Y., et al. (eds.) CIS 2005. LNCS (LNAI), vol. 3802, pp. 149–154. Springer, Heidelberg (2005). https://doi.org/10.1007/11596981_22

49. Taylor, D., Wu, T., Mavrogiannopoulos, N., Perrin, T.: RFC 5054, Using the secure remote password (SRP) protocol for TLS authentication. Technical report, RFC Editor, November 2007. https://doi.org/10.17487/rfc5054

50. Wu, T.: RFC 2944, Telnet Authentication: SRP. Technical report, RFC Editor, September 2000. https://doi.org/10.17487/rfc2944

51. Wu, T.: The secure remote password protocol. In: Proceedings of the Internet Society on Network and Distributed System Security (1998)

52. Wu, T.: The SRP Authentication and Key Exchange System, RFC 2945, September 2000

53. Wu, T.: SRP-6: Improvements and Refinements to the Secure Remote Password Protocol, October 2002

54. Zhang, M.: Analysis of the SPEKE password-authenticated key exchange protocol. IEEE Commun. Lett. **8**(1), 63–65 (2004). https://doi.org/10.1109/LCOMM.2003.822506

The Hitchhiker's Guide to Decidability and Complexity of Equivalence Properties in Security Protocols

Vincent Cheval, Steve Kremer[(✉)], and Itsaka Rakotonirina

Inria Nancy Grand-Est & LORIA, Villers-lès-Nancy, France
steve.kremer@inria.fr

Abstract. Privacy-preserving security properties in cryptographic protocols are typically modelled by observational equivalences in process calculi such as the applied pi-calculus. We survey decidability and complexity results for the automated verification of such equivalences, casting existing results in a common framework which allows for a precise comparison. This unified view, beyond providing a clearer insight on the current state of the art, allowed us to identify some variations in the statements of the decision problems—sometimes resulting in different complexity results. Additionally, we prove a couple of novel or strengthened results.

Keywords: Formal verification · Cryptographic protocols · Complexity

1 Introduction

Symbolic verification techniques for security protocols can be traced back to the seminal work of Dolev and Yao [38]. Today, after more than 30 years of active research in this field, efficient and mature tools exist, e.g. ProVerif [15] and Tamarin [50] to only name the most prominent ones. These tools are able to automatically verify full fledged models of widely deployed protocols and standards, such as TLS [14,34], Signal [29,47], the upcoming 5G standard [11], or deployed multi-factor authentication protocols [45]. We argue that the development of such efficient tools has been possible due to a large amount of more theoretical work that focuses on understanding the precise limits of decidability and the computational complexity of particular protocol classes [30,39–41,46,49].

The abovementioned results extensively cover verification for the class of *reachability* properties. Such properties are indeed sufficient to verify authentication properties and various flavors of confidentiality, even in complex scenarios with different kinds of compromise [10]. Another class of properties are

The research leading to these result has received funding from the ERC under the EU's H2020 research and innovation program (grant agreements No 645865-SPOOC), as well as from the French ANR project TECAP (ANR-17-CE39-0004-01). Itsaka Rakotonirina benefits from a Google PhD Fellowship.

V. Nigam et al. (Eds.): Scedrov Festschrift, LNCS 12300, pp. 127–145, 2020.
https://doi.org/10.1007/978-3-030-62077-6_10

indistinguishability properties. These properties express that an adversary cannot distinguish two situations and are conveniently modelled as *observational equivalences* in a cryptographic process calculus, such as the applied pi calculus. Such equivalences can indeed be used to model strong flavors of secrecy, in terms of non-interference or as a "real-or-random" experiment. Equivalences are also the tool of choice to model many other privacy-preserving properties. Such properties include anonymity [3], unlinkability properties [6,42], as well as vote privacy [37] to give a few examples. Equivalence properties are inherently more complex than reachability properties, and both the theoretical understanding and tool support are more recent and more brittle. This state of affairs triggered a large amount of recent works to increase our theoretical understanding and improve tool support.

In this paper we give an extensive overview of decidability and complexity results for several process equivalences. In particular, in this survey we give a unified view, allowing us to highlight subtle differences in the definitions of the decision problems across the literature (such as whether the term theory is part of the input or not) as well as the protocol models. Typically, models may vary in whether they allow for a bounded or unbounded number of sessions, the support of cryptographic primitives, whether they support else branches (i.e. disequality tests, rather than only equality tests), and various restrictions on non-determinism. Additionally, our technical report [25] contains full proofs of all results that are novel or that required additional arguments to make up for the differences in stating the problem compared to the original work. All the results are summarised in Table 1, and we identify several open questions. Delaune and Hirschi [36] also survey symbolic methods for verifying equivalence properties. However, they mainly discuss tool support whereas we focus on computational complexity.

2   Model

We will model protocols as processes in the applied pi calculus, and cryptographic primitives are modelled using terms equipped with rewrite rules. We assume the reader is familiar with these notions and only recall them briefly and informally. A fully-detailed model can be found in the technical report [25].

Cryptographic Primitives. As usual in symbolic protocol analysis we take an abstract view of cryptography and model the messages exchanged during the protocol as *terms* built over a set of function symbols each with a given arity. Terms are then either atomic values or function symbols applied to other terms, respecting the function's arity. Atomic values are either *constants*, i.e., function symbols of arity 0 or *names*. Constants, sometimes referred to as public names, model public values, such as agent identities or protocol tags. Names model secret values, such as keys or nonces, and are a priori unknown to the adversary. We assume an infinite set of constants and names.

Example 1. For example the encryption of a plaintext m with a key k using a symmetric encryption scheme senc is modelled by the term $\mathsf{senc}(m, k)$.   △

The functional properties of the symbols are modelled by an *equational theory*. In this work we restrict ourselves to equational theories that can be oriented into a *convergent rewriting system*. This also implies that any term t has a unique normal from $t\!\downarrow$.

Example 2. The rewrite rule $\mathsf{sdec}(\mathsf{senc}(x, y), y) \rightarrow x$ defines the behaviour of the encryption scheme: one can decrypt (apply sdec) a ciphertext $\mathsf{senc}(x, y)$ with the corresponding key y to recover the plaintext x. This behaviour is idealised by the absence of other rules for senc and sdec, modelling an assumption that no information can be extracted from a ciphertext except by possessing the decryption key. Similarly, asymmetric encryption can be modelled by the rewrite rule $\mathsf{adec}(\mathsf{aenc}(x, \mathsf{pk}(y)), y) \rightarrow x$ where pk is a symbol of arity 1 modelling public keys. Such rewrite rules can express a broad range of other primitives like pairs $(\mathsf{fst}(\langle x, y \rangle) \rightarrow x$ and $\mathsf{snd}(\langle x, y \rangle) \rightarrow y)$, hash functions (no rewrite rule) or randomised encryption (adding an additional argument to senc to explicitly represent the randomness).   △

In this survey we call a *theory* the set of non-constant function symbols together with a rewriting system. Two classes of theories are particularly important for our results. The first is the class of *subterm convergent* theories [2,12,16,22,28, 31], defined by a syntactic criterion on rewriting rules $\ell \rightarrow r$ requiring that r is either a strict subterm of ℓ or a ground term in normal form. The second is the class of *constructor-destructor* theories [16,20,22], partitioning function symbols into constructor (used to build terms) and destructors (only used in rewrite rules). In constructor-destructor theories any rewrite rule $\ell \rightarrow r$ is such that $\ell = d(t_1, \ldots, t_n)$ where d is a destructor and $t_1, \ldots, t_n, r$ do not contain any destructor. Moreover, we assume a *message* predicate $\mathsf{msg}(t)$ which holds if $u\!\downarrow$ does not contain any destructor symbol for all subterms u of t, i.e., all destructor applications in t succeeded yielding a valid message. This predicate is used to restrict to protocols that only send and accept such well-formed messages.

Protocols. Protocols are defined using processes in the applied pi calculus. The syntax of protocols is defined by the grammar of *processes*:

$$P, Q ::= \quad 0 \quad \text{if } u = v \text{ then } P \text{ else } Q \quad c(x).P \quad \bar{c}\langle u \rangle.P \quad P \mid Q$$

Intuitively the 0 models a terminated process, a conditional if $u = v$ then P else Q executes either P or Q depending on whether the terms $u\!\downarrow$ and $v\!\downarrow$ are equal, and $P \mid Q$ models two processes executed concurrently. The constructs $c(x).P$ and $\bar{c}\langle u \rangle.P$ model, respectively, inputs and outputs on a communication channel c. When the channel c is known to the attacker, e.g. when it is a constant, executing an output on c adds it to the adversary's knowledge and inputs on c are fetched from the adversary possibly forwarding a previously stored message, or computing a new message from previous outputs. Otherwise the communication is performed silently without adversarial interferences. To model an unbounded number of protocol sessions we also add the two constructs

$$P, Q ::= \quad \text{new } k.P \quad !P$$

The replication $!P$ models an unbounded number of parallel copies of P, and new $k.P$ creates a fresh name k unknown to the attacker; in particular $!$new $k.P$ models an unbounded number of sessions, each with a different fresh key. The fragment of the calculus without replication is referred to as *finite* or *bounded*. Another notable subclass is the original pi-calculus [48], referred to as the *pure* fragment, that can be retrieved with the empty theory (only names, constants and an empty rewrite system).

Semantics in an Adversarial Environment. The behaviour of processes is formalised by an operational semantics. The detailed presentation differs from one work to another [1, 21, 22] and we only give a high-level overview here. It takes the form of a transition relation $(P, \Phi) \xrightarrow{\alpha} (P', \Phi')$ on *configurations* (P, Φ) where P is the process to be executed, and Φ is called a *frame* and records the attacker knowledge. A frame is a substitution of the form $\{\text{ax}_1 \mapsto t_1, \dots, \text{ax}_n \mapsto t_n\}$ where t_i are previous outputs and ax_i are special variables called *axioms* that serve as handles to the adversary for building new terms. The label α of the transition step is called an *action* and is either

- an *unobservable action* τ which represents an internal action, such as the evaluation of a conditional or a communication on a private channel;
- an *input action* $\xi_c(\xi_t)$ where ξ_c (resp. ξ_t) represents the attacker's computation of the input's public channel (resp. of the term to be input), see *recipes* in the next section;
- an *output action* $\overline{\xi_c}\langle \text{ax}_i \rangle$ where ξ_c is again the attacker's computation of the channel, and the underlying output term is added to the frame as axiom ax_i.

We refer to the technical report [25] for full details of the semantics but provide additional intuition through the following example. Suppose that an agent S wants to send a nonce N to a recipient R. Assuming S and R already share a secret k_s, S encrypts N and k_s with the public key of R, i.e. $\text{pk}(k_R)$ where k_R is the corresponding private key. When receiving a message, R acknowledges the nonce only if the plaintext contains the shared secret. This is modelled by the following process:

$$P = S \mid R \quad \text{with} \quad S = \overline{c}\langle M \rangle \quad \text{where } M = \text{aenc}(\langle N, k_s \rangle, \text{pk}(k_R))$$
$$\text{and} \quad R = c(x). \text{ if } \text{snd}(\text{adec}(x, k_R)) = k_s \text{ then } \overline{c}\langle \text{ack} \rangle \text{ else } 0$$

with $k_s, k_R, N \in \mathcal{N}$ and $c \in \Sigma_0$. The 0 processes are omitted. The fact that the public key should be known to the attacker is modelled by a frame $\Phi_0 = \{\text{ax}_0 \mapsto \text{pk}(k_R)\}$. A "normal" execution of the protocol would be, with informal notations:

$$(S \mid R, \Phi_0) \xrightarrow{\overline{c}\langle ax_1 \rangle} (0 \mid R, \Phi_1) \qquad \text{with } \Phi_1 = \Phi_0 \cup \{ax_1 \mapsto M\} \tag{1}$$

$$\xrightarrow{c(ax_1)} (0 \mid \text{if } \text{snd}(\text{adec}(M, k_R)) = k_s \text{ then } \overline{c}\langle ack \rangle \text{ else } 0, \Phi_1) \tag{2}$$

$$\xrightarrow{\tau} (0 \mid \overline{c}\langle ack \rangle, \Phi_1) \tag{3}$$

$$\xrightarrow{\overline{c}\langle ax_2 \rangle} (0 \mid 0, \Phi_1 \cup \{ax_2 \mapsto ack\}) \tag{4}$$

Here the attacker is passive and only forwards messages. More precisely in transition (1), S sends M which is added to the frame as reference ax_1. This models that the attacker spies on the communication network and gets access to all messages sent on public channels like c. In transition (2) the attacker forwards M to R, i.e. inputs ax_1. Transition (3) is an internal test of R which leads to the final acknowledgement output (4). An active attacker would also have the capability of forging new messages and inserting them in the execution flow. For example transition (2) can be replaced by the input $\xrightarrow{c(\text{aenc}(\langle a,b \rangle, ax_0))}$ with $a, b \in \Sigma_0$: rather than forwarding M the attacker encrypts the pair of constants a, b with the public key of R (using reference ax_0) and sends it to R. In this modified execution the subsequent test would however fail. Finally let us mention that for constructor-destructor theories, all attacker-crafted terms must be valid messages, i.e. satisfy the predicate msg [22,27].

When defining security against an active attacker we quantify over all such transitions which means we consider all possible executions in an active adversarial environment. Thus even the bounded fragment yields an infinite transition system if the theory contains a non-constant function symbol (as this allows to build an unbounded number of messages).

3   Complexity of Static Equivalence (Passive Attacker)

3.1   Static Equivalence

Attacker Knowledge. As explained above, frames $\Phi = \{ax_1 \mapsto t_1, \ldots, ax_n \mapsto t_n\}$ record the outputs t_i performed during the execution of a process. They therefore enable adversarial deductions as they aggregate: for example after observing a ciphertext and the decryption key, the attacker can also obtain the plaintext by decrypting. Formally we say that one can *deduce* all terms of the form $\xi\Phi{\downarrow}$ where ξ is called a *recipe* that is a term built from function symbols, axioms $ax_i \in dom(\Phi)$, and constants. Recipes were mentioned in the previous section, in the operational semantics, as the way to specify attacker's computations. The fact that recipes cannot contain names models that names are assumed unknown to the adversary initially. For example in $\Phi = \{ax_1 \mapsto \text{senc}(t, k), ax_2 \mapsto k\}$ the term t is deducible by the recipe $\text{sdec}(ax_1, ax_2)$, regardless of k being a name (which is not allowed to occur directly in the recipe).

Indistinguishability. Some security properties against a passive attacker, i.e. a simple eavesdropper, can then be modelled as an observational equivalence of two frames: intuitively no equality test can be used to distinguish them.

For example, in a protocol that outputs a sequence of messages $t_1, \ldots, t_n$, the "real-or-random" confidentiality of a key k can be modelled as the equivalence of

$$\Phi = \{\mathsf{ax}_1 \mapsto t_1, \ldots, \mathsf{ax}_n \mapsto t_n, \mathsf{ax} \mapsto k\} \quad \Psi = \{\mathsf{ax}_1 \mapsto t_1, \ldots, \mathsf{ax}_n \mapsto t_n, \mathsf{ax} \mapsto k'\}$$

where k' is a fresh key. More formally, two frames Φ, Ψ with same domain are *statically equivalent* when for all recipes ξ_1, ξ_2 we have that

$$\xi_1 \Phi\!\downarrow = \xi_2 \Phi\!\downarrow \iff \xi_1 \Psi\!\downarrow = \xi_2 \Psi\!\downarrow .$$

In constructor-destructor theories we also require that $\mathsf{msg}(\xi_1\Phi)$ *iff* $\mathsf{msg}(\xi_1\Psi)$, modelling an assumption that the adversary can observe destructor failures.

Example 3. If k, k' are names, $\Phi = \{\mathsf{ax} \mapsto k\}$ and $\Psi = \{\mathsf{ax} \mapsto k'\}$ are equivalent, capturing the intuition that random keys cannot be distinguished. However, for the constant 0, $\Phi = \{\mathsf{ax}_1 \mapsto \mathsf{senc}(0, k), \mathsf{ax}_2 \mapsto k\}$ and $\Psi = \{\mathsf{ax}_1 \mapsto \mathsf{senc}(0, k), \mathsf{ax}_2 \mapsto k'\}$ are not equivalent since $\xi_1 = \mathsf{sdec}(\mathsf{ax}_1, \mathsf{ax}_2)$ and $\xi_2 = 0$ are equal in Φ but not in Ψ.   △

3.2   Complexity Results

We study the following decision problem referred to as STATEQ:

INPUT: A theory, two frames of same domain.
QUESTION: Are the two frames statically equivalent for this theory?

General Case. As rewriting is Turing-complete, unsurprisingly static equivalence is undecidable in general for convergent rewriting systems [2]. It is also proved in [2] that the DEDUCIBILITY problem (given a term t and a frame Φ, is t deducible in Φ?) reduces to STATEQ. As a consequence, the results of [5] imply that static equivalence is also undecidable for so-called *optimally-reducing* rewrite systems, a subclass of rewrite systems that have the finite-variant property [17].

Subterm Convergent Theories. Historically, complexity results for static equivalence only considered fixed theories [2,12], that is, the theory was not part of the input of the problem and its size was seen as a constant in the complexity analysis. This was consistent with most formalisms and verification tools at the time, which would not allow for user-defined theories and only consider a fixed set of cryptographic primitives, such as in the spi-calculus for example [4]. In particular fixed theories are considered in the following result:

Theorem 1 ([2]). *For all fixed subterm convergent theories* STATEQ *is* PTIME.

Generic PTIME-completeness results would make no sense when the theory is not part of the input, since the complexity may depend of it. Typically when using an empty theory the complexity changes:

Theorem 2 ([22]). *In the pure pi-calculus,* STATEQ *is* LOGSPACE.

However, in some sense, the PTIME bound is optimal since it is possible to provide a hardness result for a large class of fixed subterm convergent theories:

Theorem 3. *For all fixed theories containing symmetric encryption,* STATEQ *is PTIME-hard.*

Proof. (Sketch). We proceed by reduction from HORNSAT. Let X the set of variables of a Horn formula $\varphi = C_1 \wedge \ldots \wedge C_n$, and k_x names for all $x \in X \cup \{\bot\}$. Then to each clause $C = x_1, \ldots, x_n \Rightarrow x$, $x \in X \cup \{\bot\}$ we associate the term

$$t_C = \mathsf{senc}(\ldots \mathsf{senc}(\mathsf{senc}(k_x, k_{x_1}), k_{x_2}), \ldots, k_{x_n}).$$

Putting k_x under several layers of encryption ensures that k_x is deducible if all the keys $k_{x_1}, \ldots, k_{x_n}$ are deducible as well. In particular $k_\bot$ is deducible from the terms $t_{C_1}, \ldots, t_{C_n}$ *iff* the formula φ is unsatisfiable. Hence given two constants 0,1, and $\Phi = \{\mathsf{ax}_1 \mapsto t_{C_1}, \ldots, \mathsf{ax}_n \mapsto t_{C_n}\}$, we have that the frames $\Phi \cup \{\mathsf{ax} \mapsto \mathsf{senc}(0, k_\bot)\}$ and $\Phi \cup \{\mathsf{ax} \mapsto \mathsf{senc}(1, k_\bot)\}$ are statically equivalent *iff* φ is satisfiable.   □

However tools have improved since then and automated provers like KISS [28], YAPA [13] or FAST [31] are able to handle user-defined theories. It is therefore interesting today to refer to complexity analyses that account for the size of the rewrite system:

Theorem 4 ([22]). STATEQ *is coNP-complete for subterm convergent theories.*

Beyond Subterm Convergence. Although we are not aware of complexity results for the decision of static equivalence for classes larger than subterm theories, there exist decidability results. Some of the abovementioned tools, like KISS and YAPA, can actually handle most convergent rewriting system; but they naturally fail to terminate in general by undecidability of the problem. However it is proved in [28] that the termination of KISS is guaranteed for theories modelling blind signatures or trapdoor commitment (that are typically not subterm).

4   Complexity of Dynamic Equivalences (Active Attacker)

4.1   Equivalences

We expect security protocols to provide privacy against attackers that actively engage with the protocol. This can be modelled by behavioural equivalences, defining security as the indistinguishability of two instances of the protocol that differ on a privacy-sensitive attribute. There exist several candidate equivalences for modelling this notion of indistinguishability. We study two of them here and refer to [21] for details.

Trace Equivalence. The first one is *trace equivalence*. Referring to the operational semantics described in Sect. 2, we call a *trace t* a sequence of transition steps

$$t = (A_0 \xrightarrow{\alpha_1} A_1 \xrightarrow{\alpha_2} \cdots \xrightarrow{\alpha_n} A_n)$$

If tr is the word obtained after removing unobservable actions (i.e. τ's) from the word $\alpha_1 \cdots \alpha_n$, the trace t may be written $A_0 \xRightarrow{\text{tr}} A_n$ instead. Processes P_0 and P_1 are then trace equivalent when for all traces $P_i \xRightarrow{\text{tr}} (P, \Phi)$, $i \in \{0, 1\}$, there exists $P_{1-i} \xRightarrow{\text{tr}} (P', \Phi')$ such that the frames Φ and Φ' are statically equivalent. Automated verification of trace equivalence has been studied intensively for security protocols [7,20–22] and received strong tool support [17,18,23,32,33]. We refer to this problem as TRACEEQ:

INPUT: A theory, two processes.

QUESTION: Are the two processes trace equivalent?

Labelled Bisimilarity. Some other tools prove stronger equivalences, like *observational equivalence* for PROVERIF [16,19] for example. There exist several flavours of more operational bisimulation-based properties but the one usually considered in security-protocol analysis is *labelled bisimilarity* because it coincides with observational equivalence in the applied pi-calculus [1]. Formally it is an early, weak bisimulation that additionally requires static equivalence at each step; that is, it is the largest symmetric binary relation $\approx$ on processes such that $A \approx B$ implies that (1) the frames of A and B are statically equivalent, and (2) for all actions α and all transitions $A \xrightarrow{\alpha} A'$, there exists a trace $B \xRightarrow{\alpha} B'$ such that $A' \approx B'$. We refer to the following problem as BISIM:

INPUT: A theory, two processes.

QUESTION: Are the two processes labelled bisimilar?

4.2   Classical Fragments of the Calculus

In addition to the assumptions on the rewriting system (e.g. subterm convergence as in Sect. 3), there are several common restrictions made on the processes.

Conditionals and Patterns. A typical restriction on conditionals is the class of *positive* processes that only contains trivial else branches [12,21,22]. In particular for succinctness we often write $[u = v]P$ instead of if $u = v$ then P else 0.

When the rewrite system is constructor-destructor, some conditionals may also be encoded within inputs [26,27]. For that the syntax for inputs is generalised as $c(v).P$ where v is a term without destructors (but may contain variables), called a *pattern*. In terms of semantics, such inputs only accept terms that match the pattern, i.e. inputs t such that $t{\downarrow} = u\sigma{\downarrow}$ for some substitution σ, and then proceeds to execute $P\sigma$. In this paper, to ensure that protocols can be effectively implemented we require that it is possible to test with a sequence of positive conditionals that a term t matches the pattern u, and that all variables of u can be extracted by applying destructors to u. We thus define the *patterned fragment* to be the class of processes without conditionals but using pattern inputs, and where outputs do not contain destructor symbols; it is a subset of the positive fragment.

Ping Pong Protocols. These protocols [27,38,44] intuitively consist of an unbounded number of parallel processes receiving one message and sending a reply. Although the precise formalisms may differ from one work to another, the mechanisms at stake are essentially captured by processes $P = !P_1 \mid \cdots \mid !P_n$ where

$$P_i = c_i(x). [u_1^i = v_1^i] \cdots [u_{n_i}^i = v_{n_i}^i] \mathsf{new} \ k_1 \cdots \mathsf{new} \ k_{r_i}. \overline{c_i}\langle w_i \rangle$$

In particular ping-pong protocols are positive.

Simple Processes. A common middleground in terms of expressivity and decidability is the class of simple processes, for example studied in [21,26]. Intuitively, they consist of a sequence of parallel processes that operate each on a distinct, public channel—including replicated processes that generate dynamically a fresh channel for each copy. Formally they are of the form

$$P_1 \mid \cdots \mid P_m \mid !^{ch} P_{m+1} \mid \cdots \mid !^{ch} P_n \qquad \text{where } !^{ch} P = ! \ \mathsf{new} \ c_P. \overline{c_P'}\langle c_P \rangle. P$$

where each P_i does not contain any parallel operator ($\mid$) nor replication (!) and uses a unique, distinct communication channel c_{P_i}. Unlike ping pong protocols, each parallel process may input several messages and output messages that depend on several previous inputs. There exists a generalisation of simple processes called *determinate processes*, mentioned later in Sect. 5.

4.3   Complexity Results: Bounded Fragment

The bounded fragment is a common restriction to study decidability, as removing replication bounds the length of traces. However, as the attacker still has an unbounded number of possibilities for generating inputs, the transition system still has infinite branching in general. Additional restrictions are also necessary on the cryptographic primitives (as static equivalence is undecidable in general). For subterm convergent, constructor-destructor theories for example:

Theorem 5 ([22]). TraceEq *and* Bisim *are decidable in coNEXP for subterm convergent constructor-destructor theories and bounded processes.*

We do not detail the decision procedures as they are quite involved. In a nutshell, they use a dedicated constraint-solving procedure to show that, whenever trace equivalence is violated, there exists an attack trace whose attacker-input terms are at most of exponential size; in particular this shows non-equivalence to be decidable in NEXP. As before, we may also study the problem for fixed theories to investigate their influence on the complexity; typically with the empty theory:

Theorem 6 ([22]). *In the pure pi-calculus,* TraceEq *(resp.* Bisim*) is* Π_2-*complete (resp.* PSPACE*-complete) for bounded processes, and for bounded positive processes.*

However, unlike static equivalence, fixing the theory does not make it possible to obtain a bound that is better than the general one:

Theorem 7 ([22]). *There is a subterm-convergent constructor-destructor theory s.t.* TraceEq *and* Bisim *are* coNEXP*-hard for bounded positive processes.*

The theory in question [22] encodes binary trees and a couple of ad hoc functionalities. We show in the technical report [25] that, provided we discard the positivity requirement, the proof is possible with symmetric encryption and pairs only. This shows that the problem remains theoretically hard even with a minimal theory.

4.4   Complexity Results: Unbounded Case

Equivalence is undecidable in general since the calculus is Turing-complete even for simple theories. For example, Hüttel [43] shows that Minsky's two counter machines can be simulated within the spi-calculus (and hence the applied pi-calculus with symmetric encryption only). It is not difficult to adapt the proof to a simulation using only a free symbol, i.e., a function symbol h of positive arity and an empty rewrite system. These two encodings can be performed within the *finite-control fragment*, typically not Turing-complete in the pure pi-calculus (i.e. without this free function symbol) [35].

Ping Pong Protocols. While equivalence is undecidable for ping-pong protocols [27,44] some decidability results exist under additional assumptions. For example [44] studies a problem that can be described in our model essentially as Bisim for ping-pong protocols with 2 participants or less (i.e. $n \leq 2$ in the definition). This is proved decidable under some model-specific assumptions which we do not detail here. We also mention a result for patterned ping-pong protocols (cf. Sect. 4.2) without a limit on the number of participants [27]. Given a constructor-destructor theory, a ping-pong protocol P is said to be *deterministic* when each P_i (using the same notations as in the definition) can be written under the form

$$P_i = c_i(u_i).\, \text{new } k_1 \cdots \text{new } k_{r_i}.\overline{c_i}\langle v_i \rangle$$

where $u_1, \ldots, u_n$ is a family of patterns verifying:

(1) *binding uniqueness*: for all i, u_i does not contain two different variables;
(2) *pattern determinism*: for all $i \neq j$, if u_i and u_j are unifiable then $c_i \neq c_j$.

There is an additional, minor restriction on the structure of u_i and v_i that is omitted here, and we refer to [27] for details.

Theorem 8 ([27]). *For a theory limited to randomised symmetric and asymmetric encryption and digital signature,* TRACEEQ *is decidable in primitive recursive time for deterministic ping-pong protocols.*

Decidability is obtained by a reduction of the problem to the language equivalence of deterministic pushdown automata, which is decidable in primitive recursive time. A complexity lower bound for this problem is open (beyond the PTIME-hardness inherited from static equivalence, recall Theorem 3).

Simple Processes. We now study a decidability result for patterned simple processes [26]. In this work the theory is limited to symmetric encryption and pairs, and the processes must be *type compliant* and *acyclic*. Formalising the last two assumptions is quite technical and beyond the scope of this survey. We refer to the technical report [25] for more intuition and details. Compared to our other models, there is also a restriction to *atomic keys*, i.e. for all encryptions $\mathsf{senc}(u, v)$ in the process, v is either a constant, name or variable. This restriction is also enforced to attacker's recipes in the semantics by strenghtening the msg predicate.

Theorem 9 ([26]). *For a theory limited to pairs and symmetric encryption,* TRACEEQ *is coNEXP for patterned, simple, type-compliant, acyclic processes with atomic keys.*

The proof shows that equivalence of such (unbounded) processes is violated *iff* it is violated for a exponential number of sessions, and then uses a coNP decidability result in the bounded fragment [21]. However complexity was not the focus of [26] and the authors only claimed a triple exponential complexity for their procedure. Besides no lower bounds were investigated, but we proved that the problem was coNEXP-complete.

Theorem 10. *For a theory limited to pairs and symmetric encryption,* TRACEEQ *is coNEXP-hard for patterned, simple, type-compliant, acyclic processes with atomic keys.*

The reduction shares some similarities with the proof of coNEXP-hardness for trace equivalence of bounded processes (Theorem 7), compensating the more deterministic structure of simple processes by the use of replication.

Proof (Sketch). We proceed by reduction from SUCCINT 3SAT. This is a common NEXP-complete problem that, intuitively, is the equivalent of 3SAT for formulas of exponential size represented succinctly by boolean circuits. Formally a formula φ with 2^m clauses and 2^n variables $x_0, \ldots, x_{2^n - 1}$ is encoded by a boolean circuit $\Gamma : \{0,1\}^{m+2} \to \{0,1\}^{n+1}$ in the following way. If $\varphi = \bigwedge_{i=0}^{2^m - 1} \ell_i^1 \vee \ell_i^2 \vee \ell_i^3$ and $0 \leq i \leq 2^m - 1$ and $0 \leq j \leq 2$, we let x_k be the variable of the literal ℓ_i^{j+1} and b its negation bit; then $\Gamma(\bar{i}\,\bar{j}) = b\,\bar{k}$ where $\bar{i}, \bar{j}, \bar{k}$ are the respective binary representations of i, j, k. SUCCINT 3SAT is the problem of deciding, given a circuit Γ, whether the formula φ it encodes is satisfiable.

Let then φ be a formula with 2^m clauses and 2^n variables $x_0, \ldots, x_{2^n-1}$ and Γ a boolean circuit encoding this formula. We construct two simple, type-compliant, acyclic processes that are trace equivalent *iff* φ is unsatisfiable. Using pairs $\langle u, v \rangle$ we encode binary trees: a leaf is a non-pair value and, if u and v encode binary trees, $\langle u, v \rangle$ encodes the tree whose root has u and v as children. Given a term t, we build a process $P(t)$ behaving as follows:

(1) $P(t)$ first waits for an input x from the attacker. This term x is expected to be a binary tree of depth n with boolean leaves, modelling a valuation of φ (the ith leaf of x being the valuation of x_i).
(2) The goal is to make $P(t)$ verify that this valuation satisfies φ; if the verification succeeds the process outputs t. Given two constants 0 and 1, $P(0)$ and $P(1)$ will thus be trace equivalent *iff* φ is unsatisfiable.
(3) However it is not possible to hardcode within a process of polynomial size the verification that the valuation encoded by x satisfies the 2^m clauses of φ. Hence we replicate a process that, given x, verifies one clause at a time. Intuitively, the attacker will guide the verification of the 2^m clauses of φ, and whenever the ith clause has been successfully verified, the process reveals the binary representation of i (encrypted using a key unknown to the attacker).
(4) In particular, the attacker gets the encryption of all integers $0, \ldots, 2^m - 1$ only if she has successfully verified that the initial input x indeed encodes a valuation satisfying all clauses of φ. It then suffices to design a process that outputs t if the attacker is able to provide all such ciphertexts. This can be encoded by a replicated process that, upon receiving the encryption of two integers that differ only by their least significant bit, reveals the encryption of these integers with the least significant bit truncated. The verification ends when the encryption of the empty binary representation is revealed.   □

5   Variations of the Model

In this section we discuss a few variants of the model such as other notions of equivalence or different semantics of the process calculus.

Diff Equivalence. The most well-known variant of equivalence properties in security protocols is *diff-equivalence*, different variants of which are proved by the state-of-the-art PROVERIF and TAMARIN. Intuitively, diff-equivalence can be seen as an analogue of trace equivalence where two equivalent traces are also required to follow the exact same execution flow. Consider for example the processes

$$P = \overline{c}\langle u \rangle \mid \overline{c}\langle v \rangle \qquad \text{and} \qquad Q = \overline{c}\langle u' \rangle \mid \overline{c}\langle v' \rangle.$$

Given the trace of P outputting u first and then v, a proof of trace equivalence could match it with either of the two traces of Q. However diff-equivalence only considers the trace of Q outputting u' first and then v'.

Theorem 11 ([16]). *If two processes are diff-equivalent then they are also labelled bisimilar (and therefore trace equivalent).*

The converse does not hold in general, which may lead to so-called *false attacks* (non-diff-equivalent processes that are, for example, trace equivalent). Regarding decidability and complexity, we call the following problem DIFFEQ:

INPUT: A theory, two processes.

QUESTION: Are the two processes diff equivalent?

Although undecidable in general, it is known to be decidable in the bounded, positive fragment [12]. More precisely it is shown that for all *fixed* subterm convergent theories, diff-equivalence is reducible to a coNP constraint-solving problem.

Theorem 12 ([12]). *For all fixed subterm convergent theories,* DIFFEQ *is coNP for positive bounded processes.*

It is also known that DIFFEQ is coNP-hard for a theory containing only a free binary symbol h [12]. A simple proof justifies that DIFFEQ is actually coNP-hard even for the empty theory and, hence, for any fixed theory:

Theorem 13. *In the pure pi-calculus,* DIFFEQ *is coNP-complete for positive bounded processes.*

Proof. By reduction from SAT we consider a formula $\varphi = \bigwedge_{i=1}^{m} C_i$ in CNF and $\boldsymbol{x} = x_1, \ldots, x_n$ its variables. For each clause C_i, let k_i be a name and define

$$CheckSat_i(\boldsymbol{x}) = [x_{i_1} = b_{i_1}]\overline{c}\langle k_i \rangle \mid \cdots \mid [x_{i_p} = b_{i_p}]\overline{c}\langle k_i \rangle$$

where $x_{i_1}, \ldots, x_{i_p}$ are the variables of C_i and $b_{i_1}, \ldots, b_{i_p}$ their respective negation bit. That is, at least one output of k_i is reachable in $CheckSat_i(\boldsymbol{x})$ if $\boldsymbol{x}$ is a valuation of φ that satisfies C_i. In particular if we define

$$CheckSat = c(x_1). \ldots . c(x_n).(CheckSat_1(\boldsymbol{x}) \mid \cdots \mid CheckSat_m(\boldsymbol{x}))$$
$$Final(t) = c(y_1).[y_1 = k_1] \ldots c(y_m).[y_m = k_m]\ \overline{c}\langle t \rangle$$

then for two distinct constants ok, ko, we have the processes $CheckSat \mid Final(\mathsf{ok})$ and $CheckSat \mid Final(\mathsf{ko})$ diff-equivalent *iff* φ is unsatisfiable.

As far as we know the complexity of diff-equivalence has only been studied for fixed theories. However the coNP-completeness [12] can be adapted to parametric theories; inspecting the proof we observe that (1) in the complexity bounds, the dependencies in the theory are polynomial and (2) the proof uses the fact that static equivalence is PTIME for fixed theories (Theorem 1) but the arguments still hold if we only assume static equivalence to be coNP.

Theorem 14. DIFFEQ *is coNP-complete for subterm convergent theories and positive bounded processes.*

Equivalence by Session. We also briefly mention another equivalence, between diff-equivalence and trace equivalence (but incomparable with labelled bisimilarity) [24]. Known as *equivalence by session*, it was originally presented as a sound proof technique for trace equivalence in the bounded fragment, that was inducing less false attacks than diff-equivalence. We call this problem SESSEQ:

INPUT: A theory, two processes.

QUESTION: Are the two processes equivalent by session?

Surprisingly, despite practical improvements by order of magnitudes of the verification time compared to trace equivalence [24], this performance gap is not reflected in the theoretical, worst-case complexity. The same reduction as trace equivalence can indeed be used to prove equivalence by session coNEXP-hard. More details about the complexity of this problem can be found in the technical report [25].

Theorem 15. SESSEQ *is coNEXP-complete for constructor-destructor subterm convergent theories and positive bounded (resp. bounded) processes.*

The Case of Determinacy. We now mention the fragment of *determinate* processes, a generalisation of simple processes. In this fragment of the calculus, most of the studied equivalences coincide and their complexity also drops exponentially. This class has been investigated significantly [9,17,21,24] although several variants coexist in the literature, as discussed in [8]. For example the results of [9,24] hold for *action-determinate* processes, meaning that the processes never reach an intermediary state where two inputs (resp. outputs) on the same communication channel are executable in parallel; whereas a more permissive definition is used in [21]. There also exists a notion that is stricter than all of these, referred as *strong determinacy* [8]. A process is strongly determinate when (1) it does not contain private channels, (2) it is bounded, (3) all its syntactic subprocesses are strongly determinate, (4) in case the process is of the form $P \mid Q$ there exist no channels c such that both P and Q contain an input (resp. an output) on c. For example this process is action-determinate but not strongly-determinate:

$$\text{if } a = b \text{ then } c(x) \text{ else } 0 \quad \mid \quad \text{if } a = b \text{ then } 0 \text{ else } c(x).$$

Note in particular that bounded simple processes are strongly determinate.

Theorem 16 ([21,24]). *Two labelled bisimilar (resp. equivalent by session) processes are trace equivalent. The converse is true for action-determinate processes.*

In [21] it is shown that, for bounded, simple, positive processes, the equivalence problem could be reduced to the same coNP constraint-solving problem mentioned in the paragraph on diff-equivalence. Their arguments can be generalised from simple to strongly-determinate processes in a straightforward manner; however it is not clear whether this would also be true for action-determinate processes or processes with else branches. In particular we obtain for this fragment:

Table 1. Summary of the results. Colored cells indicate configurations with open problems. *All results for diff-equivalence also coincide with the results for trace equivalence, labelled bisimilarity, and equivalence by session for strongly-determinate processes.*

theory	process		STATEQ	DIFFEQ	BISIM	SESSEQ	TRACEEQ
any	bounded	any		coNP-hard	coNEXP-hard		
		pos.	coNP-complete				
any (fixed)	bounded	any	PTIME	coNP-hard	PSPACE-hard		Π₂-hard
		pos.		coNP-complete			
any	bounded	any		coNEXP coNP-hard	coNEXP-complete		
		pos.	coNP-complete				
any (fixed)	bounded	any	PTIME	coNEXP coNP-hard	coNEXP PSPACE-hard		coNEXP Π₂-hard
		pos.		coNP-complete			
sign, rsenc, raenc	unbounded	patterned, ping-pong, deterministic			PTIME-hard		PRIM REC PTIME-hard
senc, ⟨⟩	unbounded	atomic, patterned, type compliant, acyclic, simple	PTIME-complete		coNEXP-complete		
	bounded	any		coNEXP coNP-hard			
		pos.					coNEXP Π₂-hard
empty	bounded	any	LOGSPACE	coNP-complete	PSPACE-complete		Π₂-complete
		pos.					

Note: columns are STATEQ | DIFFEQ | BISIM | SESSEQ | TRACEEQ; the leftmost rotated label reads "subterm convergent" / "constructor destructor".

Theorem 17 ([21]). TRACEEQ, BISIM *and* SESSEQ *are coNP-complete for subterm convergent theories and bounded, strongly determinate, positive processes. The coNP completeness also holds for all fixed subterm convergent theories.*

Variations of the Communication Model. Although all symbolic models rely on the same fundamental ideas, several variations exist in the semantics of communication, as noted in [8]. The differences lie in the modelling of silent communications between parallel processes. In the original semantics [1], called classical in [8], communications on a same public channel between parallel processes can either be an internal, synchronous, and silent action, or be intercepted by the attacker. This is also the semantics used in the popular PROVERIF tool [16]. On the contrary, the so-called *private* semantics only allows private, unintercepted communications on channels that are unknown to the attacker, modelling an attacker that continuously eavesdrops on the network (rather than an attacker that has the capability of eavesdropping any communication). The private semantics is actually used in tools such as TAMARIN [50] and AKISS [17] and also in the presentation of equivalence by session [24].

While both semantics are equivalent when it comes to reachability properties, they surprisingly happen to be incomparable for equivalence properties [8]. All the complexity results of this paper are with respect to the private semantics. Although we did not expand on studying all the variations of complexity induced by using different semantics, most of the analyses presented in this survey are robust to these changes. Indeed, all complexity results for the bounded fragment hold for both semantics. In the unbounded case, only the private semantics has been considered in the underlying models [26, 27].

6   Summary of the Results

Table 1 summarises the results of and highlights open questions (including a few results only detailed in the technical report [25] for space reasons). Cells for which the complexity results are not tight are colored in grey. For instance, for subterm-convergent constructor-destructor theories and bounded processes, DIFFEQ is known coNEXP and coNP-hard, but the precise complexity remains unknown. We also include in this table some complexity results with the theory seen as a constant of the problem (denoted as "fixed" in the *theory* columns). The corresponding cells contain bounds applying to *all* theories of the class; e.g. for BISIM of bounded processes, with fixed subterm-convergent constructor-destructor theories, the problem is decidable in coNEXP and PSPACE-hard; despite the gap between the two bounds, they are optimal since there exist theories for which the problem is either coNEXP-hard or PSPACE. Therefore this cell is not highlighted in grey. In our opinion the most interesting open questions are:

- Can upper bounds on constructor-destructor subterm convergent theories be lifted to more general subterm convergent theories?
- Without the positivity assumption, can we tighten the complexity for diff equivalence, and strongly determinate processes?

This last question might allow to better understand why strongly determinate processes benefit from optimisations that improve verification performance that much. Finally, as witnessed by the contrast between the high complexity of equivalence by session and its practical efficiency, worst-case complexity may not always be an adequate measure.

References

1. Abadi, M., Blanchet, B., Fournet, C.: The applied pi calculus: mobile values, new names, and secure communication. J. ACM (JACM) **65**, 1–41 (2017)
2. Abadi, M., Cortier, V.: Deciding knowledge in security protocols under equational theories. Theoret. Comput. Sci. **367**, 2–32 (2006)
3. Abadi, M., Fournet, C.: Private authentication. Theoret. Comput. Sci. **322**, 427–476 (2004)
4. Abadi, M., Gordon, A.D.: A calculus for cryptographic protocols: the spi calculus. Inf. Comput. **148**, 1–70 (1999)
5. Anantharaman, S., Narendran, P., Rusinowitch, M.: Intruders with caps. In: Baader, F. (ed.) RTA 2007. LNCS, vol. 4533, pp. 20–35. Springer, Heidelberg (2007). https://doi.org/10.1007/978-3-540-73449-9_4
6. Arapinis, M., Chothia, T., Ritter, E., Ryan, M.: Analysing unlinkability and anonymity using the applied pi calculus. In: IEEE Computer Security Foundations Symposium (CSF) (2010)
7. Arapinis, M., Cortier, V., Kremer, S.: When are three voters enough for privacy properties? In: Askoxylakis, I., Ioannidis, S., Katsikas, S., Meadows, C. (eds.) ESORICS 2016. LNCS, vol. 9879, pp. 241–260. Springer, Cham (2016). https://doi.org/10.1007/978-3-319-45741-3_13
8. Babel, K., Cheval, V., Kremer, S.: On the semantics of communications when verifying equivalence properties. J. Comput. Secur. **28**(1), 71–127 (2020)
9. Baelde, D., Delaune, S., Hirschi, L.: Partial order reduction for security protocols. In: International Conference on Concurrency Theory (CONCUR) (2015)
10. Basin, D.A., Cremers, C.: Know your enemy: compromising adversaries in protocol analysis. ACM Trans. Inf. Syst. Secur. (TISSEC) **17**, 1–31 (2014)
11. Basin, D.A., Dreier, J., Hirschi, L., Radomirovic, S., Sasse, R., Stettler, V.: A formal analysis of 5G authentication. In: ACM Conference on Computer and Communications Security (CCS) (2018)
12. Baudet, M.: Sécurité des protocoles cryptographiques: aspects logiques et calculatoires. Ph.D. thesis (2007)
13. Baudet, M., Cortier, V., Delaune, S.: YAPA: a generic tool for computing intruder knowledge. ACM Trans. Comput. Log. (TOCL) **14**, 1–32 (2013)
14. Bhargavan, K., Blanchet, B., Kobeissi, N.: Verified models and reference implementations for the TLS 1.3 standard candidate. In: IEEE Symposium on Security and Privacy, (S&P) (2017)
15. Blanchet, B.: Modeling and verifying security protocols with the applied pi calculus and ProVerif. In: Foundations and Trends in Privacy and Security (2016)
16. Blanchet, B., Abadi, M., Fournet, C.: Automated verification of selected equivalences for security protocols. J. Log. Algebraic Program. **75**, 3–51 (2008)
17. Chadha, R., Cheval, V., Ciobâcă, Ş., Kremer, S.: Automated verification of equivalence properties of cryptographic protocols. ACM Trans. Comput. Log. (TOCL) **17**, 1–32 (2016)

18. Cheval, V.: APTE: an algorithm for proving trace equivalence. In: Ábrahám, E., Havelund, K. (eds.) TACAS 2014. LNCS, vol. 8413, pp. 587–592. Springer, Heidelberg (2014). https://doi.org/10.1007/978-3-642-54862-8_50

19. Cheval, V., Blanchet, B.: Proving more observational equivalences with ProVerif. In: Basin, D., Mitchell, J.C. (eds.) POST 2013. LNCS, vol. 7796, pp. 226–246. Springer, Heidelberg (2013). https://doi.org/10.1007/978-3-642-36830-1_12

20. Cheval, V., Comon-Lundh, H., Delaune, S.: Trace equivalence decision: negative tests and non-determinism. In: ACM Conference on Computer and Communications Security (CCS) (2011)

21. Cheval, V., Cortier, V., Delaune, S.: Deciding equivalence-based properties using constraint solving. Theoret. Comput. Sci. **492**, 1–39 (2013)

22. Cheval, V., Kremer, S., Rakotonirina, I.: DEEPSEC: deciding equivalence properties in security protocols theory and practice. In: IEEE Symposium on Security and Privacy (S&P) (2018)

23. Cheval, V., Kremer, S., Rakotonirina, I.: The DEEPSEC prover. In: Chockler, H., Weissenbacher, G. (eds.) CAV 2018. LNCS, vol. 10982, pp. 28–36. Springer, Cham (2018). https://doi.org/10.1007/978-3-319-96142-2_4

24. Cheval, V., Kremer, S., Rakotonirina, I.: Exploiting symmetries when proving equivalence properties for security protocols. In: ACM Conference on Computer and Communications Security (CCS) (2019)

25. Cheval, V., Kremer, S., Rakotonirina, I.: The Hitchhiker's guide to decidability and complexity of equivalence properties in security protocols (Technical report) (2020). https://hal.archives-ouvertes.fr/hal-02501577

26. Chrétien, R., Cortier, V., Delaune, S.: Decidability of trace equivalence for protocols with nonces. In: IEEE Computer Security Foundations Symposium (CSF) (2015)

27. Chrétien, R., Cortier, V., Delaune, S.: From security protocols to pushdown automata. ACM Trans. Comput. Log. (TOCL) **17**, 1–45 (2015)

28. Ciobâcă, Ş., Delaune, S., Kremer, S.: Computing knowledge in security protocols under convergent equational theories. In: Schmidt, R.A. (ed.) CADE 2009. LNCS (LNAI), vol. 5663, pp. 355–370. Springer, Heidelberg (2009). https://doi.org/10.1007/978-3-642-02959-2_27

29. Cohn-Gordon, K., Cremers, C., Garratt, L., Millican, J., Milner, K.: On ends-to-ends encryption: Asynchronous group messaging with strong security guarantees. In: ACM Conference on Computer and Communications Security (CCS) (2018)

30. Comon, H., Cortier, V.: Tree automata with one memory set constraints and cryptographic protocols. Theoret. Comput. Sci. **331**, 143–214 (2005)

31. Conchinha, B., Basin, D.A., Caleiro, C.: Fast: an efficient decision procedure for deduction and static equivalence. In: International Conference on Rewriting Techniques and Applications (RTA) (2011)

32. Cortier, V., Dallon, A., Delaune, S.: Efficiently deciding equivalence for standard primitives and phases. In: Lopez, J., Zhou, J., Soriano, M. (eds.) ESORICS 2018. LNCS, vol. 11098, pp. 491–511. Springer, Cham (2018). https://doi.org/10.1007/978-3-319-99073-6_24

33. Cortier, V., Grimm, N., Lallemand, J., Maffei, M.: A type system for privacy properties. In: ACM Conference on Computer and Communications Security (CCS) (2017)

34. Cremers, C., Horvat, M., Hoyland, J., Scott, S., van der Merwe, T.: A comprehensive symbolic analysis of TLS 1.3. In: ACM Conference on Computer and Communications Security (CCS) (2017)

35. Dam, M.: On the decidability of process equivalences for the π-calculus. Theoret. Comput. Sci. **183**, 215–228 (1997)
36. Delaune, S., Hirschi, L.: A survey of symbolic methods for establishing equivalence-based properties in cryptographic protocols. J. Log. Algebraic Methods Program. **87**, 127–144 (2017)
37. Delaune, S., Kremer, S., Ryan, M.: Verifying privacy-type properties of electronic voting protocols. J. Comput. Secur. **17**, 435–487 (2009)
38. Dolev, D., Yao, A.: On the security of public key protocols. In: Symposium on Foundations of Computer Science (FOCS) (1981)
39. Dolev, D., Even, S., Karp, R.M.: On the security of ping-pong protocols. Inf. Control **55**, 57–68 (1982)
40. Durgin, N.A., Lincoln, P., Mitchell, J.C.: Multiset rewriting and the complexity of bounded security protocols. J. Comput. Secur. **12**, 247–311 (2004)
41. Durgin, N.A., Lincoln, P., Mitchell, J.C., Scedrov, A.: Undecidability of bounded security protocols. In: Proceedings of Workshop on Formal Methods in Security Protocols (1999)
42. Filimonov, I., Horne, R., Mauw, S., Smith, Z.: Breaking unlinkability of the ICAO 9303 Standard for e-passports using bisimilarity. In: Sako, K., Schneider, S., Ryan, P.Y.A. (eds.) ESORICS 2019. LNCS, vol. 11735, pp. 577–594. Springer, Cham (2019). https://doi.org/10.1007/978-3-030-29959-0_28
43. Hüttel, H.: Deciding framed bisimilarity. Electron. Notes Theoret. Comput. Sci **68**, 1–18 (2003)
44. Hüttel, H., Srba, J.: Recursive ping-pong protocols. BRICS Report Series (2003)
45. Jacomme, C., Kremer, S.: An extensive formal analysis of multi-factor authentication protocols. In: IEEE Computer Security Foundations Symposium (CSF) (2018)
46. Kanovich, M.I., Kirigin, T.B., Nigam, V., Scedrov, A.: Bounded memory protocols. Comput. Lang. Syst. Struct. **40**(3–4), 137–154 (2014)
47. Kobeissi, N., Bhargavan, K., Blanchet, B.: Automated verification for secure messaging protocols and their implementations: a symbolic and computational approach. In: IEEE European Symposium on Security and Privacy (EuroS&P) (2017)
48. Milner, R., Parrow, J., Walker, D.: A calculus of mobile processes. I. Inf. Comput. **100**, 1–40 (1992)
49. Rusinowitch, M., Turuani, M.: Protocol insecurity with a finite number of sessions, composed keys is NP-complete. Theoret. Comput. Sci. **299**, 451–475 (2003)
50. Meier, S., Schmidt, B., Cremers, C., Basin, D.: The TAMARIN prover for the symbolic analysis of security protocols. In: Sharygina, N., Veith, H. (eds.) CAV 2013. LNCS, vol. 8044, pp. 696–701. Springer, Heidelberg (2013). https://doi.org/10.1007/978-3-642-39799-8_48

Assumption-Based Analysis
of Distance-Bounding Protocols with CPSA

Paul D. Rowe[⊠], Joshua D. Guttman, and John D. Ramsdell

The MITRE Corporation, Bedford, USA
{prowe,guttman,ramsdell}@mitre.org

Abstract. This paper, dedicated to Andre Scedrov, was inspired by conversations with him about the physical properties of distributed systems. We use CPSA, the strand space protocol analysis tool, to analyze and classify distance-bounding protocols. We introduce a model of strand spaces that explicitly accounts for physical properties like distance. We prove that non-metric, causal facts allow us to infer distance bounds. Moreover, CPSA already provides these causal conclusions about protocols. We apply this method to numerous protocols from the literature. By taking an assumption-based perspective—rather than an attack-based perspective—we introduce a taxonomy of distance-bounding protocols that compares the relative strength of different designs.

1 Introduction

A *distance-bounding protocol* is an exchange of messages between parties that include a *prover* and a *verifier* [6,16]. The verifier wants to determine whether the prover is nearby, i.e. within some application-relevant radius. This requires authenticating the prover to some extent, since generally one wants to know which party is within the radius. For instance, if a credit card is the device acting as prover, the verifier definitely needs to know what number is associated with it so that the right number will be billed.

Distance-bounding protocols have often been weak, sometimes quixotically weak. In this paper, we will approach distance-bounding protocols in three steps.

First, although the goals of a distance-bounding protocol are essentially metric—they are about how far the prover is from the verifier—we extract a non-metric model from them, using strand spaces. From this non-metric model, together with purely local metric assertions about the time elapsed for a single participant, metric consequences about space and time will follow. Lemma 2 justifies this step back to a conclusion about the distance to the prover.

Second, we show how to use the strand space protocol analyzer CPSA to extract a set of non-metric executions for each distance-bounding protocol. From these non-metric executions, we can draw conclusions about whether a protocol achieves its metric goals, with the backing of Lemma 2.

Finally, we exhibit a taxonomy classifying distance-bounding protocols by the assumptions that they require, to be sure of achieving their goals.

© Springer Nature Switzerland AG 2020
V. Nigam et al. (Eds.): Scedrov Festschrift, LNCS 12300, pp. 146–166, 2020.
https://doi.org/10.1007/978-3-030-62077-6_11

Strand Spaces in Spacetime. The idea for strand spaces came from an analogy to spacetime diagrams in physics. A spacetime diagram organizes some physical interactions by considering the *world-lines* of some entities as they progress through time, moving in space. Moreover, the entities interact through *messages*, whether transmitted as light or as other waves or particles; these messages travel no faster than the speed of light c.

Protocol analysis is structurally similar: the world-line of a principal includes message transmissions and receptions. If a principal is *regular*, i.e. acting in accordance with the protocol under study, these transmission and reception events partition into a number of *regular strands*, meaning a finite sequence of transmission and reception events $\circ \Rightarrow \circ \Rightarrow \cdots$ permitted by some protocol role. For uniformity, we divide the actions of a Dolev-Yao adversary [12] into a collection of finite sequences of transmission and reception events; these are *adversary strands*. A protocol execution consists of a finite collection of regular and adversary strands, or initial segments of them, with two main properties:

- If a reception event receives a message m, then some transmission event must have sent m, i.e. $\circ \xrightarrow{m} \circ$; and
- the finite directed graph G must be acyclic, where G's nodes $\circ$ are the events, and G's arcs $\rightarrow, \Rightarrow$ are either message communications $\rightarrow$ or the succession relation between two events along the same strand $\Rightarrow$.

These are natural properties of causality. The first says that message reception needs to be causally explained by some transmission. The second is the familiar principle that causality is well-founded: You cannot go back and encourage your grandparents to beget your parents, or not to. It certainly applies in our context, in which message transmissions and receptions occur at discrete, well-separated times, and where moreover none of the activities will stretch over long (or cosmological) timescales.

Diagrams with these two properties are *bundles*, and bundles form the strand space execution model. Bundles $\mathcal{B}$ have "forgotten" the metric that governs events in spacetime, and retained only the strand structure and communication arcs.

Each bundle $\mathcal{B}$ has a partial order $\preceq_\mathcal{B} = (\rightarrow \cup \Rightarrow)^*$, which is the weakest reflexive, transitive relation that extends the succession relation $\Rightarrow$ of nodes on the same strand, and extends the communication relation $\circ \xrightarrow{m} \circ$.

The acyclicity justifies a well-founded induction principle on $\preceq_\mathcal{B}$: If S is a non-empty set of nodes of $\mathcal{B}$, then there exist nodes in S that are $\preceq_\mathcal{B}$-minimal in S. Reasoning in strand spaces is ultimately justified by taking cases on these minimal nodes, given the permissible regular strands and adversary strands.

Hence, strand spaces are particularly natural for reasoning about distance-bounding protocols. The pure protocol analysis allows us to characterize the bundles a protocol allows. These then may be embedded in spacetime in any way that respects their causal structure, including the physical principle that causality cannot propagate faster than the speed of light. If this implies that the distance between two entities must have been below a selected bound d, then the protocol has achieved its goal.

Our overall strategy is akin to Meadows et al.'s 2007 work [24]. They capitalize on the causal characteristics of the challenge-response principles that govern security protocol correctness in general, which thereby determine how events can be ordered given the effects some of them must have on others. We add a particular realization of these principles for a model of security protocols [14]. Since that model is backed by an efficient tool, namely CPSA, we can apply the method on an industrial scale.

Mauw et al. [22] also observe the value of using the causal structure to guide protocol analysis for bounding distance. A separate source brought the problem back to our attention: Andre Scedrov and Carolyn Talcott discussed their work on distance-bounding protocols, including round-off attacks, repeatedly at the Protocol Exchange we periodically share [2,3,17,18]. The opportunity for an analysis of the kind we will present here was a consequence of those discussions, together with some preliminary work [33].

CPSA, a Cryptographic Protocol Shapes Analyzer. The protocol analysis tool CPSA implements the *enrich-by-need* method [14,28,29]. CPSA carries out protocol analysis by showing the analyst all of the minimal, essentially different executions compatible with some scenario of interest, often a very small set. By a scenario, we generally mean a situation in which some protocol roles have executed at least part way, with some assumptions that some parameters are freshly chosen, or some long-term keys are uncompromised. A *skeleton* means a formal representation of such a scenario.

Starting from a skeleton $\mathbb{A}_0$, CPSA systematically explores how to add new role instances and other information in ways that would help explain executions. CPSA does not explicitly represent adversary actions, but simply keeps track of what the adversary can obtain from the regular transmissions, subject to the assumptions. Mathematically, CPSA explores skeletons by rising in a *homomorphism ordering*, and it stops along any branch of its exploration when it runs out of possible explanations or reaches a *realized* skeleton.

A skeleton $\mathbb{B}$ is *realized* if, together with adversary actions compatible with its freshness and non-compromise assumptions, it can form a bundle $\mathcal{B}$. We say that $\mathbb{B}$ is *a skeleton of* such a bundle $\mathcal{B}$, and we say that a skeleton $\mathbb{A}$ *covers* bundle $\mathcal{B}$ if there exists a realized skeleton $\mathbb{B}$ such that $\mathbb{B}$ is a skeleton of $\mathcal{B}$, and a homomorphism $H: \mathbb{A} \rightarrow \mathbb{B}$.

The set of minimal realized skeletons are called the *shapes* for the starting skeleton $\mathbb{A}_0$. CPSA is useful because well-designed protocols often lead to small sets of shapes, even though the set of shapes is large or infinite in unfavorable cases.[1] CPSA presents the shapes in a concrete, graphical form, allowing a logically naive designer to understand the effects of varying protocol choices.

Moreover, each shape contains the events and their ordering needed for the non-metric, causal aspects of our distance-bounding analyses.

CPSA now allows assuming that certain messages pass over channels that ensure confidentiality or integrity. Any protocol implementation must discharge

---

[1] Indeed, since Andre et al. [13] proved the underlying problem class to be undecidable, uniform termination is impossible.

the assumptions, for instance by suitable cryptography. But CPSA can infer the effects of the assumptions, independent of particular choices about how to discharge them. We will use these channel assumptions in Sect. 4.

Protocol Goals as Formulas. CPSA offers a logical language $\mathcal{L}_\Pi$ to express goals for a protocol Π [15,27]. $\mathcal{L}_\Pi$ includes the following types of predicates:

- For each role $\rho \in \Pi$, and for each transmission or reception position i along ρ, $\mathcal{L}_\Pi$ contains a one-place predicate $r_{\rho,i}(n)$ that asserts that a node n is an instance of the ith event along role ρ.
- For each role $\rho \in \Pi$, and for each parameter or variable x that helps to determine ρ's instances, $\mathcal{L}_\Pi$ contains a two-place predicate $p_{\rho,x}(n,v)$ that asserts that node n's instance for the x parameter is v.
- The causal ordering $n \prec n'$ is expressed by a predicate $\mathtt{prec}(n,n')$.
- Two nodes on the same strand satisfy the *collinear* predicate $\mathtt{coll}(n,n')$.
- $\mathtt{unique}(v)$ is satisfied if v is fresh; $\mathtt{non}(k)$, if key k is non-compromised.
- Confidentiality and integrity for a channel c are $\mathtt{conf}(c)$ and $\mathtt{auth}(c)$.

Any skeleton $\mathbb{A}_0$ may be expressed by a conjunctive formula of $\mathcal{L}_\Pi$. Thus, a CPSA run starting from $\mathbb{A}_0$ determines what must be true in all Π-bundles satisfying this formula, which we call the *characteristic formula* $\mathsf{cf}(\mathbb{A}_0)$ of $\mathbb{A}_0$.

A *goal formula* is a universally quantified implication $\forall \overline{x} \,.\, \Phi \Longrightarrow \bigvee_{i \in I} \exists \overline{y}_i \,.\, \Psi_i$, where Φ and the Ψ_i are conjunctions of atomic formulas (see Definition 4).

The special case $I = \emptyset$ gives the empty disjunction $\bigvee_{i \in \emptyset}$ with no way to be true, i.e. false. A goal $\mathsf{cf}(\mathbb{A}_0) \Longrightarrow \mathsf{false}$ states that no Π-bundle exhibits the scenario $\mathbb{A}_0$. If $\mathbb{A}_0$ assumes some putative secret k is heard unprotected, expressed in a parameter predicate $p_{\mathtt{lsn},x}(n,k)$ for a special role, the conclusion false ensures non-disclosure. Formulas with non-empty conclusions express authentication properties. They say that the behavior in the hypothesis Φ requires additional behavior found in one of the conclusions Ψ_i.

Indeed, a terminating run of CPSA may be summarized as a formula, which we call a *shape analysis formula* [27]. Suppose, starting from the initial scenario $\mathbb{A}_0$, CPSA terminates with the family of shapes $\{\mathbb{B}_i\}_{i \in I}$. It has discovered the security goal formula $\mathsf{cf}(\mathbb{A}_0) \Longrightarrow \bigvee_{i \in I} \exists \overline{y}_i \,.\, \mathsf{cf}(\mathbb{B}_i)$; the homomorphisms from $\mathbb{A}_0$ to the $\mathbb{B}_i$ determine the quantified variables $\overline{y}_i$. The formula must be true because the CPSA search is *sound*, i.e. it refines any skeleton $\mathbb{A}$ to a set of skeletons that cover all of the executions that $\mathbb{A}$ covers. Moreover, it is a strongest goal with the hypothesis $\mathsf{cf}(\mathbb{A}_0)$, because each of the shapes $\mathbb{B}_i$ really is an essentially different scenario that can occur. No correct goal could rule any of them out.

Thus, the shape analysis formula is the *strongest* security goal achieved by Π for this hypothesis [15,32]. In this way, CPSA allows us to discover what security goals Π achieves, for the situations of concern to us.

As this suggests, there is a natural ordering on security goals that share the same antecedent Φ, namely the entailment ordering on their conclusions $\bigvee_{i \in I} \exists \overline{y}_i \,.\, \Psi_i$. There is also a dual ordering on security goals that share the same *conclusion* $\overline{\Psi}$. Namely, of two security goals $\Gamma_1 = \Phi_1 \Longrightarrow \overline{\Psi}$ and $\Gamma_2 = \Phi_2 \Longrightarrow \overline{\Psi}$, Γ_1 is at least as strong as Γ_2 iff Φ_2 entails Φ_1.

In Sect. 4 we use this idea to compare different protocols, according to whether their shared distance-bounding conclusions require stronger or weaker assumptions to assure. The cross-protocol use of formulas like this is justified in our work on protocol transformation [15,32].

2   Adapting the Strand Model for Distance-Bounding

Let d be the usual Euclidean distance and c be the speed of light. We add metric information to bundles in the simplest way:

Definition 1. *Let $\mathcal{B}$ be a bundle, and let $E\colon \mathsf{nodes}(\mathcal{B}) \to \mathbb{R}^4$ be a function from the nodes of $\mathcal{B}$ into spacetime. $(\mathcal{B}, E)$ is a spacetime bundle iff, for all n_1, n_2 such that $n_1 \prec_{\mathcal{B}} n_2$, letting $E(n_1) = (t_1, x_1, y_1, z_1)$, and $E(n_2) = (t_2, x_2, y_2, z_2)$:*

1. $t_1 < t_2$; *and*
2. $d((x_2, y_2, z_2), (x_1, y_1, z_1)) < c \cdot (t_2 - t_1)$.

We will write $d_E(n_2, n_1)$ for $d((x_2, y_2, z_2), (x_1, y_1, z_1))$ and $t_E(n_1)$ for t_1.

Evidently, every spacetime bundle determines a (non-metric) bundle, namely its first component. Indeed, intuitively, however the events of $\mathcal{B}$ have occurred in space and time, they will satisfy conditions 1–2.

Conversely, any bundle $\mathcal{B}$ may be embedded into spacetime, i.e. it is the first component of some spacetime bundle:

Lemma 1. *Let $\mathcal{B}$ be a bundle. There exists an $E\colon \mathsf{nodes}(\mathcal{B}) \to \mathbb{R}^4$ such that $(\mathcal{B}, E)$ is a spacetime bundle.*

Proof. We choose to let each strand be stationary. Construct E by well-founded recursion on $\preceq_{\mathcal{B}}$. For each n choose a time $t_E(n)$ that exceeds the time of its immediate predecessors enough to allow its incoming messages to arrive. There is no upper bound on the choice for $t_E(n)$. $\qquad\square$

Any skeleton $\mathbb{A}$ is compatible with or *covers* a (possibly empty) set of spacetime bundles $(\mathcal{B}, E)$, namely all those where there is a $H\colon \mathbb{A} \to \mathbb{B}$ such that $\mathbb{B}$ is a skeleton of $\mathcal{B}$.

Definition 2. *Let $\mathbb{A}$ be a skeleton with collinear nodes $n_1 \Rightarrow^+ n_2$, and let n' be a node. We say n_1, n_2 bound separation from n' in $\mathbb{A}$ iff $n_1 \preceq_{\mathbb{A}} n' \preceq_{\mathbb{A}} n_2$.*

Lemma 2. *Let $(\mathcal{B}, E)$ be a spacetime bundle; $H\colon \mathbb{A} \to \mathbb{B}$, and $\mathbb{B}$ be a skeleton of $\mathcal{B}$. If n_1, n_2 bound separation from n' in $\mathbb{A}$, then*

$$d_E(H(n_1), H(n')) + d_E(H(n'), H(n_2)) < c \cdot (t_E(H(n_2)) - t_E(H(n_1))).$$

That is, using a local clock along the strand of $H(n_1)$, the principal executing it can bound the distance to the node $H(n')$. Thus, reasoning about ordering in a skeleton gives a uniform way to bound the distance between corresponding events in all the spacetime bundles it covers.

Proof. The homomorphism H preserves the ordering relations, as does the embedding of the realized skeleton $\mathbb{B}$ into the bundle $\mathcal{B}$. Thus, condition 2 in Definition 1 yields the desired inequality.   □

We express requirements on distance-bounding protocols as security goals. Since we must talk about particular formulas and free variables, we will write formal variables $\lceil n \rceil$ with a ceiling in the next few paragraphs to distinguish them from our informal variables n ranging over nodes. Subsequently, we will revert to the usual ambiguity between mentioning formal variables and using informal variables. To express bounded separation goals, we distinguish particular formal variables $\lceil n_1, n_2, n' \rceil$.

Definition 3. *Let Γ be a security goal $\forall \overline{x} . \ \Phi \implies \bigvee_{\ell \in L} \exists \overline{y}_\ell . \ \Psi_\ell$ in $\mathcal{L}_\Pi$ with non-empty L; let $\lceil n_1, n_2 \rceil$ be node variables among the variables $\overline{x}$, and $\lceil n' \rceil$ be among the variables $\overline{y}_\ell$ for every ℓ. Then:*

1. $\Gamma, \lceil n_1, n_2, n' \rceil$ *is a* distance-bounding requirement *for protocol Π (or a* requirement, *for short).*
2. Π achieves *the requirement $\Gamma, \lceil n_1, n_2, n' \rceil$, iff, for every realized Π-skeleton $\mathbb{B}$ and each variable assignment η such that η satisfies $\mathbb{B} \models_\eta \Phi$, there is some $\ell \in L$ and an η' extending η such that $\mathbb{B} \models_{\eta'} \Psi_\ell$ and moreover $\eta'(\lceil n_1 \rceil), \eta'(\lceil n_2 \rceil)$ bound separation from $\eta'(\lceil n' \rceil)$ in $\mathbb{B}$.*

Since both the conclusions Ψ_ℓ and bounding separation are preserved by homomorphisms, as soon as they are satisfied in a branch of a CPSA, they will remain true thereafter. Moreover, by Lemma 2, the requirement ensures that some strand satisfying Ψ_ℓ will be no farther from a strand satisfying Φ than the locally elapsed time $\Delta t \cdot c/2$ between the ith and j^{th} node.

Hence, suppose we want to check if Π achieves a requirement $\Gamma, \lceil n_1, n_2, n' \rceil$, where Γ is of the form $\forall \overline{x} . \ \Phi \implies \bigvee_{\ell \in L} \exists \overline{y}_\ell . \ \Psi_\ell$, and Φ is the characteristic formula of a CPSA starting scenario $\mathbb{A}_0$, i.e. $\Phi = \mathsf{cf}(\mathbb{A}_0)$.

1. Execute CPSA starting from the scenario $\mathbb{A}_0$, obtaining the set of shapes $\{H_\ell \colon \mathbb{A}_0 \to \mathbb{B}_\ell\}_{\ell \in I}$;
2. ascertain that each $\mathbb{B}_\ell \models_\eta \Gamma$;
3. for the satisfying variable assignments η', check that $\eta'(\lceil n_1 \rceil), \eta'(\lceil n_2 \rceil)$ bound separation from $\eta'(\lceil n' \rceil)$ in $\mathbb{B}_\ell$.

In the favorable case in which I is finite, these steps terminate.

We can easily express bounded separation as a conjunctive formula in the variables $\lceil n_1, n_2, n' \rceil$, namely:

$$\mathsf{prec}(\lceil n_1 \rceil, \lceil n' \rceil) \wedge \mathsf{prec}(\lceil n' \rceil, \lceil n_2 \rceil),$$

which we will denote $\mathsf{bnd\_sep}(\lceil n_1 \rceil, \lceil n_2 \rceil, \lceil n' \rceil)$. Thus, in practice we perform surgery on the given goal Γ to obtain Γ^+:

$$\forall \overline{x} . \ \Phi \implies \bigvee_{\ell \in L} \exists \overline{y}_\ell . \ (\Psi_\ell \wedge \mathsf{bnd\_sep}(\lceil n_1 \rceil, \lceil n_2 \rceil, \lceil n' \rceil)).$$

CPSA can check this security goal directly, as we illustrate in the next section.

3 Examples

In this section, we show how CPSA is used to find and fix a flaw in the Terrorist-fraud Resistant and Extractor-free Anonymous Distance-bounding (TREAD) protocol [4]. The aim of its authors is to "obtain provable terrorist-fraud resistant protocols without assuming that provers have any long-term secret key". Alas, the case in which TREAD is implemented with public key cryptography as shown in Fig. 2 of [4] has an authentication failure.

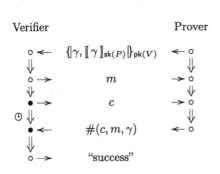

Verifier   Prover

Fig. 1. TREAD Protocol

Figure 1 shows our model of the TREAD protocol. Each participant, V and P, has a public key $\mathsf{pk}(\cdot)$ and a private key $\mathsf{sk}(\cdot)$. A message is encrypted with $\{\!| \cdot |\!\}_{\mathsf{pk}(\cdot)}$ and signed with $[\![\cdot]\!]_{\mathsf{sk}(\cdot)}$. The first message exchanged in the protocol is γ signed by the prover and then encrypted for the verifier.

All distance-bounding protocols include a *fast phase*, where one principal measures the time it takes for a sequence of message interactions. Our modeling of TREAD abstracts away details of its fast phase by a pair of messages. The two bullets • near the clock ⊙ in the Verifier role show the beginning and end of the timed fast phase.

In TREAD, γ is a pair of random n-bit values $\gamma = (\alpha, \beta)$. During the fast phase of the protocol, the Verifier sends n one bit messages that make up the contents of randomly chosen n-bit message c. The Prover responds to the reception of c_i with r_i, where

$$r_i = \begin{cases} \alpha_i & \text{if } c_i = 0, \\ \beta_i \oplus m_i & \text{if } c_i = 1. \end{cases}$$

The Verifier declares success if it receives the responses it expects within the protocol's time bound. If the adversary cannot obtain γ, the adversary is highly unlikely to provide the right n values for the r_i. In that case, n bounded separation claims are likely to hold.

In our protocol representation with a single fast exchange, the Prover sends the hash of c, m, and γ, and the Verifier declares success if it receives that message. Thus, in our version, we would like bounded separation to hold where n_1, n_2 are the two Verifier nodes on the timed edge, and n' is the Prover node that transmits $\#(c, m, \gamma)$. The security goal Γ asserts that if a Verifier run completes, a Prover run with matching V, P, γ, m, c parameters should also complete.

Analysis of TREAD. Figure 1 describes the TREAD protocol. Consider the point-of-view in which the Verifier has run to completion with freshly chosen m, c and non-compromised $\mathsf{sk}(P), \mathsf{sk}(V)$. What else must have happened?

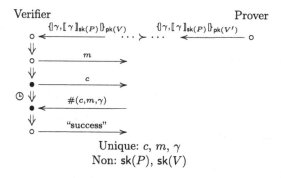

Fig. 2. TREAD Shape

The shape found by CPSA is displayed in Fig. 2. CPSA infers that the Prover was active, but it may only have transmitted its first message, which may have been altered before delivery by the adversary. The message received at the Verifier's 4th node can be synthesized by the adversary. CPSA is telling us that there are bundles that are compatible with the shape in which adversary strands synthesize all the messages received by the Verifier using only the message sent at the Prover's first (and only) node. Thus, neither Γ nor the bounded separation property holds.

CPSA explains each step it takes on its way to finding its answers, and a knowledgeable user can use this information to fix the protocol. However, we press on, trying to fix the problem by adding a confirming message at the end of the protocol. This is a reasonable thing to try, as at least one industrial protocol uses this technique to (slowly) authenticate the replies sent during the fast phase [8]. Plus, it is intuitively clear that this should allow the Verifier to conclude that the Prover must have engaged in a fast phase.

Figure 3 shows the amended protocol we call TREAD+. When started with the point-of-view skeleton in which the Verifier runs to completion, CPSA finds the shape in Fig. 3. This time, CPSA concludes that the Prover must have run to full length. However, the mismatch between the first message sent by the Prover and received by the Verifier is still present, so Γ fails. Message $\#(c, m, \gamma)$ is received by the Verifier on the second timed node, and is also sent by the Prover. However, CPSA does not report that the transmission has to precede the reception: The adversary can synthesize $\#(c, m, \gamma)$ before the Prover sends it! This occurs because the Prover's random values leak in the first message. Thus, bounded separation again fails.

The RETREAD protocol fixes the authentication problem in the TREAD protocol. It alters the first message by including the name of the Verifier, V, in the signed part of the message. Therefore, the first message in both roles of the protocol is $\{|\gamma, [\![\gamma, V]\!]_{\mathsf{sk}(P)}|\}_{\mathsf{pk}(V)}$. When CPSA is started with the point-of-view skeleton in which the Verifier runs to completion, it finds the shape in Fig. 4. No adversary behavior need occur in bundles compatible with this shape. What CPSA learns is expressed in the shape analysis sentence:

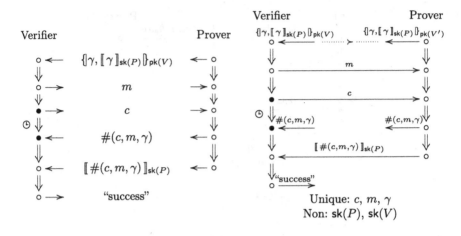

Fig. 3. TREAD+ Protocol (l) and relevant shape (r)

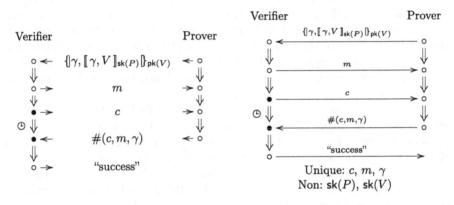

Fig. 4. The RETREAD protocol (l) and its shape (r)

If a Verifier with parameters P, V, c, m, and γ runs to completion, and
- c, m, and γ are assumed to be uniquely originating, and
- $\mathsf{sk}(P)$ and $\mathsf{sk}(V)$ are assumed to be non-originating,

then a Prover with parameters P, V, c, m, and γ ran to completion, with bounded separation for the timed Verifier nodes and the 3th Prover node.

Adding V's name inside the signature in the Prover's last message in TREAD+ also forces V and P to agree on V's identity, ensuring γ remains secret and ensuring bounded separation. However, RETREAD is a superior protocol since it is shorter and requires only a single signature.

4   Taxonomy

Much of the recent literature on symbolic analysis of distance-bounding protocols has focused on classifying protocols according to their ability to resist various kinds of attacks (e.g. [7,9,22]). Our position is that it is more useful to categorize protocols according the security goals they achieve. We follow the approach from [32] in which security goals are expressed as first-order logical formulas. The strength ordering of goal formulas is naturally captured by logical implication. If Γ_1 and Γ_2 are security goals, then Γ_1 is at least as strong as Γ_2 iff $\Gamma_1 \Rightarrow \Gamma_2$.

Definition 4. *A security goal is a closed formula* $\Gamma \in \mathcal{L}_\Pi$ *of the form*

$$\forall \bar{x} . (\Phi \Longrightarrow \bigvee_{k \in K} \exists \bar{y}_k . \Psi_k)$$

where Φ *and* Ψ *are conjunctions of atomic formulas. We write* $\mathsf{hyp}(\Gamma) = \Phi$ *and* $\mathsf{conc}(\Gamma) = \bigvee_{k \in K} \exists \bar{y}_k . \Psi_k$.

Fundamentally, all distance-bounding protocols have the same minimal goal. If the verifier accepts a run apparently with prover P, then P must have responded to the challenge after the start of the fast phase of the protocol and before its completion. Protocols may have more stringent authentication requirements such as needing the prover to agree on the verifier's name and other authenticated data. But often such agreement is achieved in the service of the main goal which is to ensure P must be close. We can naturally express this in our goal language.

To say that the verifier has accepted a run apparently with P we may write

$$\Phi_1(n, P) = \mathtt{VerifierDone}(n) \wedge \mathtt{Prover}(n, P)$$

where $\mathtt{VerifierDone}(\cdot)$ is a predicate that holds for the last node of a verifier's run, and $\mathtt{Prover}(\cdot)$ signifies the verifier's value for the prover's identity.

To express the requirement that P respond to the verifier's challenge during the fast phase, we need to identify the nodes starting and stopping the fast phase. We can write

$$\Phi_2(n_1, n_2) = \mathtt{StartTimer}(n_1) \wedge \mathtt{StopTimer}(n_2) \wedge \mathtt{coll}(n_1, n_2)$$

where $\mathtt{StartTimer}(\cdot)$ and $\mathtt{StopTimer}(\cdot)$ serve to identify the nodes starting and stopping the fast phase on the verifier's strand. $\mathtt{coll}(n_1, n_2)$ states that these nodes start and stop the fast phase on the same strand. We similarly must ensure that we are referring to the fast phase of the same strand as the one accepting the run with P. It suffices to express that n and n_1 are on the same strand. Putting it all together, we have:

$$\Phi(n, P, n_1, n_2) = \Phi_1(n, P) \wedge \Phi_2(n_1, n_2) \wedge \mathtt{coll}(n, n_1) \tag{1}$$

Equation 1 serves as the hypothesis for the distance-bounding security goal. The conclusion must state that P responded to the challenge during the fast phase. As all distance-bounding protocols have an event in which the prover sends the reply to a challenge, we use $\mathtt{ProverReply}(\cdot)$ to denote such a node of a prover strand. We again use $\mathtt{Prover}(\cdot, \cdot)$ to express that the prover's identity for the `ProverReply` node is P. Finally, we use `bnd_sep` to express the ordering required for the fast phase. The result is:

$$\Psi(n_1, n_2, n', P) = \texttt{ProverReply}(n') \wedge \texttt{Prover}(n', P) \wedge \texttt{bnd\_sep}(n_1, n_2, n') \quad (2)$$

The basic distance-bounding security goal is thus:

$$\mathsf{DB} = \forall n, P, n_1, n_2 . \Phi(n, P, n_1, n_2) \implies \exists n' . \Psi(n_1, n_2, n', P) \quad (3)$$

However, no protocol can achieve DB as formulated in Eq. 3. The assumptions in hyp(DB) are too weak to imply bounded separation. First, distance bounding is hopeless unless the verifier chooses fresh values. We will henceforth always assume this, adopting a corresponding strengthening Φ' in place of Eq. 1.

But also, Φ makes no assumption about the authenticity or confidentiality of any communications channels—either directly or through assumptions on cryptographic keys. It is well-known that authentic or confidential channels cannot be constructed without access to an authentic or confidential channel [21], or corresponding secret keys. We identify a collection of additional assumptions that can help to ensure a protocol can achieve the goal of bounding the distance of the apparent prover. We identify three main types of assumptions:

s. Secrecy of long-term keys (private keys and/or shared symmetric keys)
f. Freshness of prover-chosen values
a. Authenticity of messages received during the fast phase

The assumption that a given long-term key has been kept secret is familiar for cryptographic protocols of all types. In surveying the literature, there are typically three types of long-term keys that distance-bounding protocols tend to rely on: private keys belonging to the prover ($\mathsf{sk}(P)$), private keys belonging to the verifier ($\mathsf{sk}(V)$), and symmetric keys shared by the prover and verifier ($\mathsf{ltk}(P, V)$). We may state the corresponding secrecy requirements as follows:

$$s_{\mathsf{ltk}}(n, P, V) = \texttt{Verifier}(n, V) \wedge \texttt{Prover}(n, P) \wedge \texttt{non}(\mathsf{ltk}(P, V)) \quad (4)$$

$$s_{\mathsf{prv}}(n, P) = \texttt{Prover}(n, P) \wedge \texttt{non}(\mathsf{sk}(P)) \quad (5)$$

$$s_{\mathsf{vrf}}(n, V) = \texttt{Verifier}(n, V) \wedge \texttt{non}(\mathsf{sk}(V)) \quad (6)$$

The freshness of prover-chosen values can also play an important role in the success of distance-bounding protocols. If a verifier believes the prover's nonces to be randomly chosen and shared only with the verifier, then by incorporating the nonces into the reply during the fast phase the verifier can conclude it really is the prover providing the reply. However, there are several natural reasons this assumption may not be justified. In many distance-bounding protocols the prover has very limited computational power, and so may also not have a reliable source of randomness. It may also be the case that a dishonest and distant prover is willing to share their nonces with a malicious accomplice who is in physical proximity with the verifier. This is related to Terrorist Fraud Attacks [11], about which we say more in a later section. We may state the freshness assumption on a prover's nonce as follows:

$$f(n, np) = \texttt{ProverNonce}(n, np) \wedge \texttt{uniq}(np) \tag{7}$$

In some protocols (e.g. TREAD), the prover contributes two nonces. In those cases, for each of the nonces, f will include a pair of conjuncts like Eq. 7. In our analysis below, we always make the same assumption on all of the prover's nonces. That is, we either assume all nonces are fresh, or we don't assume any are.

Finally, some protocols might be run in environments where it is reasonable to assume that only regular (i.e. honest) provers can provide the replies during the fast phase. Consider, for example, a secure facility that enforces physical access control to a building that uses a distance-bounding protocol to gate access to special rooms. The fast phase may use near-field communication meaning that only those provers who have already passed the initial access control would be within range. This is one way to ensure the authenticity of messages received during the fast phase.

Whether malicious parties have access to the timed channel inbound to the verifier is related to Distance Hijacking Attacks [9]. We may state the assumption that the inbound timed channel is authentic as follows:

$$a(n, timed) = \texttt{TimedChannel}(n, timed) \wedge \texttt{auth}(timed) \tag{8}$$

Equations 4–8 allow us to define a family of distance-bounding security goals according to which subsets of the assumptions we include. Many protocols only require the prover to have access to a single long-term key. Depending on whether it is a shared symmetric key or a private signing key, we would use either Eq. 4 or 5. By making all possible combinations of assumptions of type $s, f,$ and a, we naturally generate eight possible goals which we denote $DB^{\mathcal{P}(\{sfa\})}$ according to which subset of $\{s, f, a\}$ is included in the assumptions of $hyp(DB^{\mathcal{P}(\{sfa\})})$ So, for example, $DB = DB^{\emptyset}$ because we make none of the assumptions. The goal that only assumes authenticity of messages received during the fast phase is denoted DB^a, (with the set braces suppressed for readability) which stands for the formula:

$$\forall n, P, n_1, n_2, timed \ . \ \Phi(n, P, n_1, n_2) \wedge a(n, timed) \implies \exists n' \ . \ \Psi(n_1, n_2, n', P).$$

Following the ideas in [32], this family of goals is naturally ordered by implication. Goal formulas that make fewer assumptions are naturally stronger. Figure 5 depicts the ordering of the family $DB^{\mathcal{P}(\{sfa\})}$. Only the superscripts are denoted in the diagram. This partial ordering can serve as a yard stick to measure the relative strength of a variety of designs for distance-bounding protocols. If a protocol satisfies the goal at one point in the partial order, then it satisfies all goals below it (since they are ordered by implication). Therefore, we can evaluate protocols according to the maximal goals they achieve. We performed a survey of numerous protocols from the literature, and the boxes indicate the maximal strength achieved by at least one of the protocols we studied.

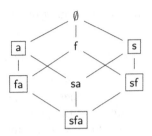

Fig. 5. Strength ordering for $DB^{\mathcal{P}(\{sfa\})}$. Boxes indicate maximal strength achieved by at least one protocol from our survey.

For protocols such as TREAD that rely on two long-term keys, one of the keys is typically the prover's signing key, while the other is either the verifier's private decryption key ($sk(V)$) or a shared symmetric key ($ltk(P, V)$). We may wish to separate the assumptions we make about their secrecy. This yields a bigger family of goals denoted $DB^{ss'fa}$ where s represents the assumption s_{prv} (Eq. 5), and s' represents either s_{ltk} (Eq. 4) or s_{vrf} (Eq. 6) depending on the design of the protocol. This yields a bigger lattice of security goals depicted in Fig. 6. Again, the boxes indicate maximal strengths achieved by at least one protocol among those we surveyed.

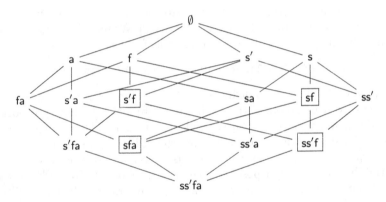

Fig. 6. Strength ordering for $DB^{\mathcal{P}(\{ss'fa\})}$. Boxes indicate maximal strength achieved by at least one protocol from our survey.

As stated above, we analyzed numerous protocols from the literature. Our intent is not to be exhaustive, but rather to demonstrate the utility of assumption-based analyses for comparing the relative strength of different designs of distance-bounding protocols. Space constraints preclude an exhaustive description of all the analyses, but the results are summarized in Table 1[2] and we discuss a few noteworthy highlights below. The times reported are based on runs using a 2018 MacBook Air with 1.6 GHz Dual-Core Intel Core i5 processor with 16 GB of RAM. They represent the total elapsed time after verifying all 8 or 16 variants of the goal depending on how many long-term keys the protocol uses. As the table makes clear, CPSA is an extremely efficient tool.

---

[2] Cf. https://github.com/mitre/cpsaexp/tree/master/doc/dist_bnd_prots.

Table 1. Various distance-bounding protocols ordered by strength.

Protocol	Strength	Elapsed time (s)
Protocols with a single long-term key		
Hancke and Kuhn [16]	$\{s\}, \{a\}$	0.03
Kim and Avoine [19]	$\{s\}, \{a\}$	0.03
Munilla et al. [26]	$\{s\}, \{a\}$	0.07
Reid et al. [31]	$\{s\}, \{a\}$	0.04
Swiss-Knife [20]	$\{s\}, \{a\}$	0.05
Mauw et al. [23]	$\{sf\}, \{a\}$	0.03
Meadows et al. [24]	$\{sf\}, \{fa\}$	0.05
BC-Signature [6]	$\{sf\}$	0.06
CRCS [30]	$\{sf\}$	0.06
BC-FiatShamir [6]	$\{sfa\}$	0.10
Protocols with two long-term keys		
Paysafe [8]	$\{sf\}, \{s'f\}$	0.12
TREAD-SK [5]	$\{sfa\}, \{s'f\}$	0.08
TREAD-PK [5]	$\{sfa\}$	0.05
TREAD variants introduced in this paper		
TREAD-SK+	$\{sfa\}, \{s'f\}$	0.16
RETREAD-SK	$\{sfa\}, \{s'f\}$	0.07
RETREAD-PK	$\{sfa\}, \{ss'f\}$	0.06
TREAD-PK+	$\{sfa\}$	0.16

We first note that it seems to be easy for distance-bounding protocols to satisfy the weakest goal (DB^{sfa} or $DB^{ss'fa}$). Every protocol we checked was secure under the strongest set of assumptions. The weakest protocol we discovered was Brands and Chaum's early adaptation of the Fiat-Shamir identification scheme into a distance-bounding scheme [6]. This weak result may not be entirely accurate, but might rather be an artifact of modeling algebraic properties with logical axioms.

At the other end of the spectrum, there is a collection of protocols that all satisfy both DB^a and DB^s which are incomparable goals [16, 19, 20, 26, 31]. Indeed, this is the best we can hope for. As we have already seen, $DB^\emptyset$ is impossible to achieve due to the need to have access to at least one confidential or authentic channel [21]. DB^f is unsatisfiable for the same reason. Therefore, simultaneously satisfying both "shoulders" of Fig. 5 is the maximum strength possible.

It is instructive to consider the design principles used by various protocols that contribute to their strength or weakness. The family of protocols achieving the maximum strength are all based on the same core design. Namely, in the fast phase, the prover's reply cryptographically binds the verifier's challenge with a long-term shared symmetric key that serves to authenticate the prover.

The exact way in which these values are cryptographically bound varies widely, but it typically involves generating a hash of a message containing at least the shared key and the verifier's nonce. Much of the variation in the designs is attributable to the need to make the cryptographic operation as simple as possible. More computationally intensive operations force the verifier to accept longer threshold times for the round trip because the verifier needs to account for the computation time as well. Longer threshold times generally provide weaker distance guarantees. Our symbolic analysis only ensures that an upper bound on the distance can be achieved. Since it does not consider the computation time explicitly, CPSA does not distinguish among these protocols.

In order to better understand how different designs fall short of the maximum strength, consider the protocol from Mauw et al. [23]. Rather than creating an explicit binding between the long-term shared symmetric key and the verifier's nonce, they create an implicit binding. They do this with a message in the setup phase. During this first phase, the prover sends a nonce to the verifier encrypted under their long-term, shared symmetric key. During the fast phase, the prover combines the verifier's nonce with its own nonce from the first phase. This creates an implicit binding between the verifier's nonce and the long-term key. But, crucially, this binding only succeeds if the long-term key has not been compromised, *and* the prover's nonce is indeed random and fresh. An adversary near to the verifier who is capable of guessing the prover's nonce (or an adversary who can coerce the prover into leaking its nonce during the first phase) can cause a distant prover to appear close to the verifier. Thus, in this protocol, the security of the long-term key is not enough. The verifier must also assume the prover's nonce is not available outside the bounds of the protocol execution. This is why it does not achieve DB^s, but does achieve DB^{sf}. When considering the goal DB^a, the authenticity of the fast channel guarantees the prover is honest. Since the reply also contains the prover's identity, this authentic channel successfully authenticates the prover's identity.

The remaining protocols suffer in similar ways. Generally speaking, they also perform implicit bindings between the verifier's nonce and the long-term key. In attempting to make the prover's response as fast as possible to compute, various techniques are chosen that suffer from subtle algebraic collisions. For example, the bindings are frequently created by performing an xor operation which is very efficient. But such values are not inherently integrity protected, so there is an opportunity for algebraic manipulation. This is contrast to a standard hash function which may be slower to compute, but which does not admit such algebraic manipulations.

We also analyzed TREAD together with a shared-key version also present in [5]. We distinguish them as TREAD-PK and TREAD-SK respectively. We can now compare them to the altered versions we introduced in Sect. 3. Our earlier analysis focused solely on the goal $DB^{ss'f}$. While TREAD-PK fails to satisfy that goal, it does satisfy DB^{sfa}. This says that, when the verifier assumes its timed inbound channel is authentic, the verifier need not rely on the secrecy of its own private key to achieve the distance bound. TREAD-SK additionally satisfies $DB^{s'f}$

which implies the goal we investigated in Sect. 3 but is also incomparable with the single strongest goal achieved by TREAD-PK. Although TREAD-SK does not satisfy any goals stronger than the strongest one achieved by TREAD-PK, it does satisfy goals the public key version does not. In this sense, TREAD-SK is strictly stronger than TREAD-PK.

Notice that TREAD-PK+ provides no benefit beyond TREAD-PK, and similarly for the shared-key versions. RETREAD-PK, on the other hand, is slightly stronger than TREAD-PK, although not quite as strong even as TREAD-SK since it doesn't achieve $DB^{s'f}$. RETREAD-SK satisfies the same goals as TREAD-SK, so adds no value in a shared key context.

5 Related Work

Spacetime vs. Causality. A key aspect of our approach to analyzing distance-bounding protocols is the lack of any explicit account of time or distance in the protocol models. We are not the first to make the observation that a causality-based analysis is informative enough to draw conclusions about time and distance. We follow the ideas taken by Mauw et al. in [22] in which they introduce a semantic model that explicitly accounts for time and distance. They then relate that model to the execution model underlying Tamarin [25] just as we relate spacetime bundles and realized skeletons. The underlying Dolev-Yao adversary model of Tamarin is sufficient to capture all relevant attacks.

This is in contrast to the work of Chothia et al. [7], which uses specific classes of processes to model dishonest provers, instead of relying solely on the underlying Dolev-Yao processes. Their approach is also causality-based: Rather than model time and distance quantitatively, they model "places;" processes in the same place are nearby. They adapt the pi-calculus operational semantics [1] so that processes that communicate during the fast phase have the same place.

Various other symbolic approaches account for time and distance more explicitly [9,10,17]. Our core insight arose from discussions with Andre Scedrov and Carolyn Talcott when they presented their work starting with [17] at our Protocol Exchange meeting. Our discussions suggested we could separate the analysis of the causal constraints from an analysis of the quantitative constraints of time and distance. We believed we could first reason causally and collect quantitative constraints along the way. The causal reasoning would justify deriving a tolerable strand-local delay from the desired quantitative distance bound. Lemmas 1–2 justify the procedure.

Attack-Based vs. Assumption-Based. Since the very early days of studying distance-bounding protocols the focus has been on preventing various types of attacks. The attacks are commonly referred to by names such as Terrorist Fraud and Mafia Fraud. One frequently finds intuitive definitions of these attack types based on the relative locations of Dolev-Yao attackers, honest & dishonest provers, and the verifier. The informality of these intuitions can make it quite difficult to interpret how they should be formally defined in any given model.

Indeed, as observed in [22], there still remains some disagreement around the appropriate formal definition of Terrorist Fraud.

The clearest, most formal definitions we could find were in [7]. They introduce systematic definitions of dishonest provers for Mafia Fraud and Terrorist Fraud. They then explore the set of combinations of verifiers, Dolev-Yao attackers, and honest & dishonest provers in all possible relative locations, and organize them into a hierarchy of attacks.

In our view, focusing on and explicitly modeling different attack types runs counter to the spirit of most modern approaches to symbolic analysis of protocols. The community no longer makes distinctions about whether an adversary executes a reflection attack or a replay attack. The community no longer creates different protocol models for closed systems and open systems.

Our approach is based on a lesson learned by one of the authors from an observation made by Andre Scedrov regarding the classic Needham-Schroeder protocol. The standard view that Lowe found a previously undiscovered attack is somewhat misleading. The original protocol was secure *under the assumptions* made by Needham and Schroeder. Lowe's attack did not invalidate old security claims. It merely showed that the desired authentication property doesn't hold *under a weaker set of assumptions*. In the language of strand spaces, Needham and Schroeder assumed that initiators only engage in sessions with responders whose private keys are non-originating. Indeed, under such an assumption the protocol *does* achieve the desired conclusion. The question of whether such an assumption is justified is separate from that of whether the protocol achieves the right conclusion under the assumption.

This observation motivates the assumption-driven analysis in contrast to an attack-driven analysis. We believe our focus on altering assumptions instead of altering attacks helps focus attention on the security goals achieved by a distance-bounding protocol regardless of the type of attack. Of course, the assumptions one makes are closely related to the types of attacks considered. But we find the shift in perspective to be enlightening.

Nevertheless, we have a conjecture connecting our lattice of assumptions to the standard attack types in the literature. Figure 7 annotates our assumption lattice for protocols using one long-term key with three attack types: Mafia Fraud (MF), Terrorist Fraud (TF), and Distance Hijacking (DH). For each attack type, we associate it with the weakest goal such that, if the protocol achieves that goal, then it resists the given attack type. So, for instance, if a protocol achieves DB^a then it resists distance hijacking attacks. This association is only an informal conjecture at this point. The corresponding association for protocols

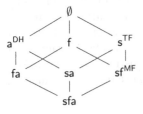

Fig. 7. Conjecture: attacks and the assumptions that prevent them.

with two long-term keys is less clear. It is worth noting, however, the relative order of the attacks in Fig. 7 matches the corresponding order embedding in the attack hierarchy of [7]. Establishing this conjecture, or making it more precise, would require a more careful comparison of the semantics of the formal models.

Symbolic vs. Quantitative Analysis. Our symbolic analysis relies on a simple and clear use of causal structures to infer the security goals achieved by distance-bounding protocols. However, the simplicity and clarity is often obtained by abstracting away the finer details of the fast phase. The fast phase typically involves repeated round trips of single-bit messages, which we represent as a single round trip of many-bit messages. The causal structure arises out of unique-origination assumptions on nonces which preclude any other agent from being able to send the nonce without first receiving it.

However, at the bit level, any given round trip does not guarantee the desired causal order because an adversary or a dishonest prover always has a chance of guessing the correct reply bit before receiving the challenge bit. The causal conclusions only emerge probabilistically over time as challenge-response round trips are performed. Symbolic analyses are therefore incapable of yielding insights about how many challenges a verifier should issue to be confident of the causal consequences. Recent work by Andre Scedrov and others explicitly addresses this question for the Hancke-Kuhn family of distance-bounding protocols [2,3].

Another creative line of inquiry by Scedrov and others [18] involves a more nuanced analysis of just how strongly the timing constraints can bound the distance between the verifier and the prover. They introduce a model that accounts not only for the time it takes for a message to travel through space, but also for the time it takes for instructions to execute. Because low-powered processors can often only perform one instruction during any given clock tick, there can be time between the event of starting the timer and the event of sending the challenge that is unaccounted for by the timing constraint. They discover the possibility of an "Attack Between the Ticks" in which a distant prover takes advantage of this time discrepancy to appear much closer than they actually are.

6   Conclusion

In this paper we introduced a version of strand spaces that explicitly accounts for the physical properties of spacetime. We demonstrated that it is always possible to embed the standard strand space bundles into spacetime bundles in such a way that any quantitative constraints on distance and time are satisfied.

This justifies using CPSA without modifications to analyze the security of distance-bounding protocols, illustrated by analyzing and repairing the TREAD protocol which had previously been shown to be vulnerable to attack. A survey of various distance-bounding protocols from the literature places them in a taxonomy of protocols according to their strength. In contrast to the prevailing trend, we organize our taxonomy not on the basis of attacks that are possible, but—dually—on the basis of the assumptions required for a verifier to bound the distance to a given prover.

We believe the shift in perspective to an assumption-based taxonomy from an attack-based one provides a clearer understanding of the conditions under which distance-bounding protocols succeed and fail.

References

1. Abadi, M., Fournet, C.: Mobile values, new names, and secure communication. In: 28th ACM Symposium on Principles of Programming Languages (POPL 2001), pp. 104–115 (2001)
2. AlTurki, M.A., Kanovich, M.I., Kirigin, T.B., Nigam, V., Scedrov, A., Talcott, C.L.: Statistical model checking of distance fraud attacks on the hancke-kuhn family of protocols. In: Lie, D., Mannan, M. (eds) Proceedings of the 2018 Workshop on Cyber-Physical Systems Security and PrivaCy, CPS-SPC@CCS 2018, Toronto, ON, Canada, 19 October 2018, pp. 60–71. ACM (2018)
3. Alturki, M.A., Ban Kirigin, T., Kanovich, M., Nigam, V., Scedrov, A., Talcott, C.: A multiset rewriting model for specifying and verifying timing aspects of security protocols. In: Guttman, J.D., Landwehr, C.E., Meseguer, J., Pavlovic, D. (eds.) Foundations of Security, Protocols, and Equational Reasoning. LNCS, vol. 11565, pp. 192–213. Springer, Cham (2019). https://doi.org/10.1007/978-3-030-19052-1_13
4. Avoine, G., et al.: A terrorist-fraud resistant and extractor-free anonymous distance-bounding protocol. In: Karri, R., Sinanoglu, O., Sadeghi, A.-R., Yi, X. (eds.) Proceedings of the 2017 ACM on Asia Conference on Computer and Communications Security, AsiaCCS 2017, Abu Dhabi, United Arab Emirates, 2–6 April 2017, pp. 800–814. ACM (2017)
5. Avoine, G., et al.: A terrorist-fraud resistant and extractor-free anonymous distance-bounding protocol. IACR Cryptology ePrint Archive 2017, 297 (2017)
6. Brands, S., Chaum, D.: Distance-bounding protocols. In: Helleseth, T. (ed.) EUROCRYPT 1993. LNCS, vol. 765, pp. 344–359. Springer, Heidelberg (1994). https://doi.org/10.1007/3-540-48285-7_30
7. Chothia, T., de Ruiter, J., Smyth, B.: Modelling and analysis of a hierarchy of distance bounding attacks. In: Enck, W., Felt, A.P. (eds.) 27th USENIX Security Symposium, USENIX Security 2018, Baltimore, MD, USA, 15–17 August 2018, pp. 1563–1580. USENIX Association (2018)
8. Chothia, T., Garcia, F.D., de Ruiter, J., van den Breekel, J., Thompson, M.: Relay cost bounding for contactless EMV payments. In: Böhme, R., Okamoto, T. (eds.) FC 2015. LNCS, vol. 8975, pp. 189–206. Springer, Heidelberg (2015). https://doi.org/10.1007/978-3-662-47854-7_11
9. Cremers, C.J.F., Rasmussen, K.B., Schmidt, B., Čapkun, S.: Distance hijacking attacks on distance bounding protocols. In: IEEE Symposium on Security and Privacy, SP 2012, San Francisco, California, USA, 21–23 May 2012, pp. 113–127. IEEE Computer Society (2012)
10. Debant, A., Delaune, S.: Symbolic verification of distance bounding protocols. In: Nielson, F., Sands, D. (eds.) POST 2019. LNCS, vol. 11426, pp. 149–174. Springer, Cham (2019). https://doi.org/10.1007/978-3-030-17138-4_7
11. Desmedt, Y.: Major security problems with the 'unforgeable' (feige)-fiat-shamir proofs of identity and how to overcome them. In: SECURICOM 1988, pp. 15–17 (1988)

12. Dolev, D., Yao, A.: On the security of public-key protocols. IEEE Trans. Inf. Theory **29**, 198–208 (1983)

13. Durgin, N., Lincoln, P., Mitchell, J., Scedrov, A.: Multiset rewriting and the complexity of bounded security protocols. J. Comput. Secur. **12**(2), 247–311 (2004). Initial version appeared in Workshop on Formal Methods and Security Protocols, 1999

14. Guttman, J.D.: Shapes: surveying crypto protocol runs. In: Cortier, V., Kremer, S. (eds.) Formal Models and Techniques for Analyzing Security Protocols, Cryptology and Information Security Series. IOS Press (2011)

15. Guttman, J.D.: Establishing and preserving protocol security goals. J. Comput. Secur. **22**(2), 201–267 (2014)

16. Hancke, G.P., Kuhn, M.G.: An RFID distance bounding protocol. In: First International Conference on Security and Privacy for Emerging Areas in Communications Networks, SecureComm 2005, Athens, Greece, 5–9 September 2005, pp. 67–73. IEEE (2005)

17. Kanovich, M., Ban Kirigin, T., Nigam, V., Scedrov, A., Talcott, C.: Timed multiset rewriting and the verification of time-sensitive distributed systems. In: Fränzle, M., Markey, N. (eds.) FORMATS 2016. LNCS, vol. 9884, pp. 228–244. Springer, Cham (2016). https://doi.org/10.1007/978-3-319-44878-7_14

18. Kanovich, M.I., Kirigin, T.B., Nigam, V., Scedrov, A., Talcott, C.L.: Time, computational complexity, and probability in the analysis of distance-bounding protocols. J. Comput. Secur. **25**(6), 585–630 (2017)

19. Kim, C.H., Avoine, G.: RFID distance bounding protocol with mixed challenges to prevent relay attacks. In: Garay, J.A., Miyaji, A., Otsuka, A. (eds.) CANS 2009. LNCS, vol. 5888, pp. 119–133. Springer, Heidelberg (2009). https://doi.org/10.1007/978-3-642-10433-6_9

20. Kim, C.H., Avoine, G., Koeune, F., Standaert, F.-X., Pereira, O.: The swiss-knife RFID distance bounding protocol. In: Lee, P.J., Cheon, J.H. (eds.) ICISC 2008. LNCS, vol. 5461, pp. 98–115. Springer, Heidelberg (2009). https://doi.org/10.1007/978-3-642-00730-9_7

21. Maurer, U.M., Schmid, P.E.: A calculus for security bootstrapping in distributed systems. J. Comput. Secur. **4**(1), 55–80 (1996)

22. Mauw, S., Smith, Z., Toro-Pozo, J., Trujillo-Rasua, R.: Distance-bounding protocols: Verification without time and location. In: 2018 IEEE Symposium on Security and Privacy, SP 2018, Proceedings, 21–23 May 2018, San Francisco, California, USA, pp. 549–566. IEEE Computer Society (2018)

23. Mauw, S., Smith, Z., Toro-Pozo, J., Trujillo-Rasua, R.: Post-collusion security and distance bounding. In: Cavallaro, L., Kinder, J., Wang, X., Katz, J. (eds.) Proceedings of the 2019 ACM SIGSAC Conference on Computer and Communications Security, CCS 2019, London, UK, 11–15 November 2019, pp. 941–958. ACM (2019)

24. Meadows, C.A., Poovendran, R., Pavlovic, D., Chang, L., Syverson, P.F.: Distance bounding protocols: authentication logic analysis and collusion attacks. In: Poovendran, R., Roy, S., Wang, C. (eds.) Secure Localization and Time Synchronization for Wireless Sensor and Ad Hoc Networks. Advances in Information Security, vol. 30, pp. 279–298. Springer, Heidelberg (2007). https://doi.org/10.1007/978-0-387-46276-9_12

25. Meier, S., Schmidt, B., Cremers, C., Basin, D.: The TAMARIN prover for the symbolic analysis of security protocols. In: Sharygina, N., Veith, H. (eds.) CAV 2013. LNCS, vol. 8044, pp. 696–701. Springer, Heidelberg (2013). https://doi.org/10.1007/978-3-642-39799-8_48

26. Munilla, J., Peinado, A.: Distance bounding protocols for RFID enhanced by using void-challenges and analysis in noisy channels. Wirel. Commun. Mobile Comput. 8(9), 1227–1232 (2008)

27. Ramsdell, J.D.: Deducing security goals from shape analysis sentences. The MITRE Corporation (2012). http://arxiv.org/abs/1204.0480

28. Ramsdell, J.D., Guttman, J.D.: CPSA4: A cryptographic protocol shapes analyzer (2017). https://github.com/mitre/cpsaexp

29. Ramsdell, J.D., Guttman, J.D., Liskov, M.D., Rowe, P.D.: The CPSA Specification: A Reduction System for Searching for Shapes in Cryptographic Protocols. The MITRE Corporation (2009). http://hackage.haskell.org/package/cpsa. source distribution, doc directory

30. Rasmussen, K.B., Capkun, S.: Realization of RF distance bounding. In: 19th USENIX Security Symposium, Washington, DC, USA, 11–13 August 2010, Proceedings, pp. 389–402. USENIX Association (2010)

31. Reid, J., Nieto, J.M.G., Tang, T., Senadji, B.: Detecting relay attacks with timing-based protocols. In: Bao, F., Miller, S. (eds.) Proceedings of the 2007 ACM Symposium on Information, Computer and Communications Security, ASIACCS 2007, Singapore, 20–22 March 2007, pp. 204–213. ACM (2007)

32. Rowe, P.D., Guttman, J.D., Liskov, M.D.: Measuring protocol strength with security goals. Int. J. Inf. Secur. 15(6), 575–596 (2016). https://doi.org/10.1007/s10207-016-0319-z. http://web.cs.wpi.edu/~guttman/pubs/ijis_measuring-security.pdf

33. Thayer, F.J., Swarup, V., Guttman, J.D.: Metric strand spaces for locale authentication protocols. In: Nishigaki, M., Jøsang, A., Murayama, Y., Marsh, S. (eds.) IFIPTM 2010. IAICT, vol. 321, pp. 79–94. Springer, Heidelberg (2010). https://doi.org/10.1007/978-3-642-13446-3_6

Modelchecking Safety Properties in Randomized Security Protocols

Matthew S. Bauer[1], Rohit Chadha[2(✉)], and Mahesh Viswanathan[3]

[1] Galois Inc., Portland, USA
[2] University of Missouri, Columbia , USA
chadhar@missouri.edu
[3] University of Illinois, Urbana-Champaign, USA

Abstract. Automated reasoning tools for security protocols model protocols as non-deterministic processes that communicate through a Dolev-Yao attacker. There are, however, a large class of protocols whose correctness relies on an explicit ability to model and reason about randomness. Although such protocols lie at the heart of many widely adopted systems for anonymous communication, they have so-far eluded automated verification techniques. We propose an algorithm for reasoning about safety properties for randomized protocols. The algorithm is implemented as an extension of Stochastic Protocol ANalyzer (SPAN), the mechanized tool that reasons about the indistinguishability properties of randomized protocols. Using SPAN, we conduct the first automated verification on several randomized security protocols and uncover previously unknown design weaknesses in several of the protocols we analyzed.

1 Introduction

As security protocols are vulnerable to design flaws, machine-aided formal analysis is often utilized to verify their security guarantees. Such analysis must be carried out in the presence of an attacker that can read, intercept, modify and replay all messages on public channels, and potentially send its messages. The presence of the attacker makes the analysis challenging. In order to aid automation, the analysis is often carried out in the so-called *Dolev-Yao* model where messages are modeled as terms in a first-order vocabulary, the assumption of perfect cryptography is made. In the Dolev-Yao model, the attacker controls all communication, non-deterministically schedule the participants, and non-deterministically inject new messages, which are computed using the whole communication transcript.

Until recently, verification techniques in this domain have converged around modeling and verifying protocols that are purely non-deterministic, where non-determinism is used to model concurrency as well as the interaction between protocol participants and their environment. In this setting, decades of work have produced many sophisticated analysis tools [5,11,15,30,45]. There are, however, a large class of protocols whose correctness depends on an explicit ability to model and reason about randomness. With privacy goals in mind,

© Springer Nature Switzerland AG 2020
V. Nigam et al. (Eds.): Scedrov Festschrift, LNCS 12300, pp. 167–183, 2020.
https://doi.org/10.1007/978-3-030-62077-6_12

these protocols lie at the heart of many anonymity systems such as Crowds [41], mix-networks [22], onion routers [34] and Tor [29]. Cryptographic protocols also employ randomness to achieve fair exchange [10,31], vote privacy in electronic voting [4,21,42,44] and denial of service prevention [37]. The formal verification of this class of protocols has thus-far received little systematic attention.

In the absence of a systematic framework, there have been primarily two approaches to verify randomized security protocols. Works such as [49] use probabilistic model checkers [27,39] to reason about probabilistic behavior in systems like Crowds. These ad-hoc techniques fail to capture the Dolev-Yao attacker in full generality and do not provide a general verification framework. Other works in the symbolic model [28,38] simply abstract away essential protocol components that utilize randomization, such as anonymous channels. By making these simplifying assumptions, such analysis may miss key attacks. Indeed, we discovered in our analysis an attack on the FOO electronic voting protocol [32] that has long served as a key benchmark in the analysis of anonymity properties in the Dolev-Yao model. Our attack emerges by realizing the perfectly anonymous channels in the FOO by threshold-mixes and was missed by previous analysis.[1]

The critical challenge in the formal verification of randomized security protocols is the subtle interaction between non-determinism and randomization. If the attacker can base its non-deterministic computation on the results of private coin tosses of the participants, then the analysis necessarily may yield false attacks in correct protocols (see examples in [13,16,19,23,33]). Thus, the attacker behavior should be restricted to perform the same *computation* in any two protocol executions whose communication transcripts are indistinguishable to it. This observation is at the heart of the first framework to analyze randomized security protocols proposed in [9,17,43]. In this framework, the indistinguishability of two traces is captured by the trace-equivalence from the applied π-calculus [2]. The first-of-its-kind model-checking tool Stochastic Protocol ANalyzer (SPAN) for checking the indistinguishability of two protocols in this framework was presented in [8]. SPAN was used to verify the 3-ballot electronic voting protocol [44] in [8].

Contributions. In this work, we describe an algorithm for analyzing the reachability-based safety properties of randomized protocols that were implemented as an extension of SPAN. The algorithm follows the bounded model checking approach of the equivalence checking in SPAN and assumes that the attacker sends messages of bounded size. The problem of checking safety reduces to the problem of computing reachability of acyclic finite state Partially-Observable Markov Decision Processes (POMPDs). The analysis of finite POMDPs is, in general undecidable. However, since we deal with acyclic POMDPs, the problem of checking reachability is decidable and can be computed by converting the POMDP into a fully-observable belief Markov Decision Processes. Our algorithm exploits the acyclicity of the POMDPs to construct

---

[1] A similar attack was also discovered by hand in [6] where the analysis of FOO protocol is carried out in the computational model.

the belief MDP on-the-fly by discovering the states of the belief MDP using the Depth-First-Search strategy that is often used to solve graph reachability problems.

We use SPAN to conduct the first automated symbolic analysis of several protocols including mix-networks [22], the FOO electronic voting protocol [32] and Prêt à Voter [42]. Our analysis shows that realizing perfectly anonymous channels in the FOO protocol requires non-trivial modification to the protocol design, which if not done carefully, can lead to errors. In addition, a bug in the design of the Prêt à Voter protocol was uncovered (see Sect. 2.2). In order to fix the bug, we propose computing the cyclic offsets in the construction of Prêt à Voter using psuedorandom permutations instead of hash functions.

Related Work. Modeling cryptographic protocols in a process calculus allowing operations for both non-deterministic and probabilistic choice was first proposed in [36]. Unfortunately, the calculus did not capture many important properties of the threat model, such as the ability for protocol participants to make private coin tosses. As a result, properties of these processes are required to be formulated through a notion of bisimulation too strong to capture many natural properties. The calculus upon which our techniques are built first appeared in [9], where the authors studied the conditions under which reachability properties of randomized security protocols are preserved by composition. In [43] the composition framework was extended to handle equivalence properties. SPAN was originally presented in [8], which discusses the design and implementation of the algorithms for checking equivalence properties. For randomized security protocols, the complexity of verifying reachability and equivalence properties was studied in [17]. The material presented here also appears in the Ph.D. thesis of Matthew S. Bauer (See [7]), and we refer the reader to the thesis for a detailed discussion of the tool architecture and of experimental results.

2 Randomized Security Protocols

In what follows, we give the details behind several security protocols that utilize randomization. These protocols will serve as running examples upon which we demonstrate how our techniques can be used for modeling and automated analysis.

2.1 Mix Networks

A mix-network [22] is a routing protocol used to break the link between a message's sender and receiver. The unlinking is achieved by routing messages through a series of proxy servers, called mixes. Each mix collects a batch of encrypted messages, privately decrypts each message, and forwards the resulting messages in random order. More formally, consider a sender Alice (A) who wishes to send a message m to Bob (B) through mix (M). Alice prepares a cipher-text of the form

$$\mathsf{aenc}(\mathsf{aenc}(m, n_1, \mathsf{pk}(B)), n_0, \mathsf{pk}(M))$$

where aenc is asymmetric encryption, n_0, n_1 are nonces and $\mathsf{pk}(M)$, $\mathsf{pk}(B)$ are the public keys of the Mix and Bob, respectively. Upon receiving a batch of N such cipher-texts, the mix M unwraps the outer layer of encryption on each message using its secret key and then randomly permutes and forwards the messages. A passive attacker, who observes all traffic but does not otherwise modify the network, cannot (with high probability) correlate messages entering and exiting the mix M. Unfortunately, this simple design, known as a *threshold mix*, is vulnerable to a straightforward active attack. To expose Alice as the sender of the message $\mathsf{aenc}(m, n_1, \mathsf{pk}(B))$, an attacker forwards Alice's message along with $N-1$ dummy messages to the mix M. In this way, the attacker can distinguish which of M's N output messages is not a dummy message and hence must have originated from Alice. Although active attacks of this nature cannot be thwarted completely, several mix-network designs have been proposed to increase the overhead associated with carrying out such an attack.

2.2 Prêt à Voter

Prêt à Voter [42] is a mix-network based voting protocol that provides a simple and intuitive mechanism by which a set of voters $(V_1, ..., V_n)$ can carry out elections with the help of a set of honest tellers $(T_1, ..., T_k)$ and an honest election authority (A). Each teller has two public key pairs. Using these keys and a set of random values, the authority creates a set of ballot forms with the following properties. Each ballot has two columns; the left column lists the candidates in a permuted order and the right column provides space for a vote to be recorded. The bottom of the right column also holds an "onion" which encodes the permuted ordering (cyclic offset) for the candidates on the left-hand side of the ballot.

The precise construction of a ballot is as follows. The authority first generates a random seed,

$$\text{seed} := g_0, g_1, ..., g_{2k-1}$$

where each g_i (for $i \in \{1, ..., 2k - 1\}$), called a germ, is drawn from an appropriately sized field. For a candidate list of size v, the seed is used generate the cyclic offset

$$\theta := \sum_{i=0}^{2k-1} d_i (\text{mod } v)$$

where $d_i := \mathsf{hash}(g_i)(\text{mod } v)$. Each teller i has public keys $\mathsf{pk}(T_{2i})$ and $\mathsf{pk}(T_{2i-1})$ which are used to construct the onion

$$\{\langle g_{2k-1}, \{\langle g_{2k-1}, ... \{\langle g_0, D_0\rangle\}_{\mathsf{pk}(T_0)} ...\rangle\}_{\mathsf{pk}(T_{2k-2})}\rangle\}_{\mathsf{pk}(T_{2k-1})}$$

where D_0 is a nonce uniquely chosen for each onion. Each layer $D_{i+1} := \{\langle g_i, D_i\rangle\}_{\mathsf{pk}(T_i)}$ asymmetrically encrypts a germ and the previous layer of the onion.

The election authority generates a number of ballots which far exceed the number of voters. In order to cast a vote, a voter authenticates with the authority, after which a random ballot is chosen by the voter. In the voting booth, the voter marks his/her choice on the right-hand side of the ballot and removes the left-hand side for shredding. The values on the right side of the ballot (the vote position and onion) are read by a voting device and then retained by the voter as a receipt. Once read by the voting device, the values are passed to the tellers that manipulate pairs of the form $\langle r_{2i}, D_{2i} \rangle$. The first teller receives the pair $\langle r, D_{2k} \rangle$ where r is the vote position, and D_{2k} is the onion. Upon receiving such a pair, each teller T_{i-1} performs the following operations.

- Apply the secret key $\mathsf{sk}(T_{2i-1})$ to D_{2i} to reveal the germ g_{2i-1} and the next layer of the onion D_{2i-1}.
- Recover $d_{2i-1} = \mathsf{hash}(g_{2i-1})(\mathsf{mod}\ v)$ and obtain $r_{2i-1} = (r_{2i} - d_{2i-1})(\mathsf{mod}\ v)$.
- Form the new pair $\langle r_{2i-1}, D_{2i-1} \rangle$.

After applying this transformation for each pair in the batch it receives, teller T_{i-1} performs a secret shuffle on the resulting transformed pairs. Teller T_{i-1} then repeats this process on the shuffled values using its second secret key $\mathsf{sk}(T_{2i-2})$ to obtain a new set of pairs with the form $\langle r_{2i-2}, D_{2i-2} \rangle$. These pairs are shuffled again and then passed to the next teller T_{i-2}. The output of the last teller is the value of r_0 which identifies a voter's vote.

Our analysis of this version of the Prêt à Voter protocol has uncovered a previously unknown flaw in the protocol's design. The error arises from the assumption that the elements of the field from which the germs are drawn are evenly distributed when their hash is taken modulo v. To understand this error in more detail, let us consider the simple case when there are two candidates (0 and 1) and one teller. Let F be a field with M elements and

$$F_j = \{g \mid g \in F \text{ and } \mathsf{hash}(g)(\mathsf{mod}\ 2) = j\}$$

for $j \in \{0, 1\}$. There is no guarantee that $F_0 = F_1$ and thus the probability of the two cyclic offsets $\theta_0 = (\frac{F_0}{F})(\frac{F_0}{F}) + (\frac{F_1}{F})(\frac{F_1}{F})$ and $\theta_1 = 2(\frac{F_0}{F})(\frac{F_1}{F})$ in the randomly chosen ballots may be different. This can give an attacker an advantage in attempting to infer a vote from a ballot receipt: the attacker will guess that cyclic shift is the one happens with higher probability. To fix this issue, the hash function should be replaced by a pseudo-random permutation.

3   Randomized Applied π-Calculus

In this section, we present our core process calculus for modeling cryptographic protocols with coin tosses. The presentation of the calculus is borrowed from [8], and closely resembles the ones from [9,17,43]. As was first proposed in [36], it extends the applied π-calculus by the inclusion of a new operator for probabilistic choice.

3.1 Terms, Equational Theories and Frames

A signature $\mathcal{F}$ contains a finite set of function symbols, each with an associated arity and two special countable sets of constant symbols $\mathcal{M}$ and $\mathcal{N}$ representing public and private names, respectively. Variable symbols are the union of two disjoint sets $\mathcal{X}$ and $\mathcal{X}_w$, used to represent protocol and frame variables, respectively. The sets $\mathcal{F}$, $\mathcal{M}$, $\mathcal{N}$, $\mathcal{X}$ and $\mathcal{X}_w$ are required to be pairwise disjoint. Terms are built by the application of function symbols to variables and terms in the standard way. Given a signature $\mathcal{F}$ and $\mathcal{Y} \subseteq \mathcal{X} \cup \mathcal{X}_w$, we use $\mathcal{T}(\mathcal{F}, \mathcal{Y})$ to denote the set of terms built over $\mathcal{F}$ and $\mathcal{Y}$. The set of variables occurring in a term u is denoted by $\mathsf{vars}(u)$. A ground term is one that contains no free variables. The *depth* of a term t is defined to be the depth of the dag that represents t.

A substitution σ is a partial function with a finite domain that maps variables to terms, where $\mathsf{dom}(\sigma)$ will denote the domain and $\mathsf{ran}(\sigma)$ will denote the range. For a substitution σ with $\mathsf{dom}(\sigma) = \{x_1, \ldots, x_k\}$, we will denote σ as $\{x_1 \mapsto \sigma(x_1), \ldots, x_k \mapsto \sigma(x_k)\}$. A substitution σ is said to be ground if every term in $\mathsf{ran}(\sigma)$ is ground and a substitution with an empty domain will be denoted as $\emptyset$. Substitutions can be extended to terms in the usual way and we write $t\sigma$ for the term obtained by applying the substitution σ to the term t.

Our process algebra is parameterized by an equational theory $(\mathcal{F}, E)$, where $\mathcal{F}$ is a signature and E is a set of $\mathcal{F}$-Equations. By an $\mathcal{F}$-Equation, we mean a pair $u = v$ where $u, v \in \mathcal{T}(\mathcal{F} \setminus \mathcal{N}, \mathcal{X})$ are terms that do not contain private names.

Example 1. We can model primitives for symmetric encryption/decryption and a hash function using the equational theory $(\mathcal{F}_{\mathsf{senc}}, E_{\mathsf{senc}})$ with signature $\mathcal{F}_{\mathsf{senc}} = \{\mathsf{senc}/2, \mathsf{sdec}/2, \mathsf{h}/1\}$ and equations $E_{\mathsf{senc}} = \{\mathsf{sdec}(\mathsf{senc}(m, k), k) = m\}$.

Two terms u and v are said to be equal with respect to an equational theory $(\mathcal{F}, E)$, denoted $u =_E v$, if $E \vdash u = v$ in the first order theory of equality. For equational theories defined in the preceding manner, if two terms containing private names are equivalent, they will remain equivalent when the names are replaced by arbitrary terms. We often identify an equational theory $(\mathcal{F}, E)$ by E when the signature is clear from the context. An equational theory E is said to be trivial if $u =_E v$ for any terms u and v and, otherwise it is said to be non-trivial. For the remainder of this work, we will assume equational theories are non-trivial. Processes are executed in an environment that consists of a frame $\varphi : \mathcal{X}_w \to \mathcal{T}(\mathcal{F})$ and a binding substitution $\sigma : \mathcal{X} \to \mathcal{T}(\mathcal{F})$.

Definition 1. *Two frames φ_1 and φ_2 are said to be statically equivalent in equational theory E, denoted $\varphi_1 \equiv_E \varphi_2$, if $\mathsf{dom}(\varphi_1) = \mathsf{dom}(\varphi_2)$ and for all $r_1, r_2 \in \mathcal{T}(\mathcal{F} \setminus \mathcal{N}, \mathcal{X}_w)$ we have $r_1\varphi_1 =_E r_2\varphi_1$ iff $r_1\varphi_2 =_E r_2\varphi_2$.*

Intuitively, two frames are statically equivalent if an attacker cannot distinguish between the information they contain. A term $u \in \mathcal{T}(\mathcal{F})$ is deducible from a frame φ with recipe $r \in \mathcal{T}(\mathcal{F} \setminus \mathcal{N}, \mathsf{dom}(\varphi))$ in equational theory E, denoted $\varphi \vdash_E^r u$, if $r\varphi =_E u$. We often omit r and E and write $\varphi \vdash u$ if they are clear from the context.

3.2   Process Syntax

We assume a countably infinite set of labels $\mathcal{L}$ and an equivalence relation $\sim$ on $\mathcal{L}$ that induces a countably infinite set of equivalence classes. For $\ell \in \mathcal{L}$, $[\ell]$ denotes the equivalence class of ℓ. Each equivalence class is assumed to contain a countably infinite set of labels. Operators in our grammar will come with a unique label from $\mathcal{L}$, which, together with the relation $\sim$, will be used to mask the information an attacker can obtain about the actions of a process. When an action with label ℓ is executed, the attacker will only be able to infer $[\ell]$.

Processes in our calculus are a finite parallel composition of roles, which intuitively are used to model a single actor in a system/protocol. Please note that we are modeling only a finite number of sessions. Hence we do not allow replication in our protocol syntax. Roles, in turn, are constructed by combining atomic actions through sequential composition and probabilistic choice. Formally, an atomic action is derived from the grammar

$$A := 0 \,\big|\, \nu x^\ell \,\big|\, (x := u)^\ell \,\big|\, [c_1 \wedge \ldots \wedge c_k]^\ell \,\big|\, \mathsf{in}(x)^\ell \,\big|\, \mathsf{out}(u)^\ell$$

where $\ell \in \mathcal{L}$, $x \in \mathcal{X}$ and $c_i \in \{\top, u = v\}$ for all $i \in \{1, \ldots, k\}$ where $u, v \in \mathcal{T}(\mathcal{F} \setminus \mathcal{N}, \mathcal{X})$. In the case of the assignment rule $(x := u)^\ell$, we additionally require that $x \notin \mathsf{vars}(u)$. A role is derived from the grammar

$$R := A \,\big|\, (R \cdot R) \,\big|\, (R +_p^\ell R)$$

where $p \in [0, 1]$, $\ell \in \mathcal{L}$ and $x \in \mathcal{X}$. The 0 process does nothing. The process νx^ℓ creates a fresh name and binds it to x while $(x := u)^\ell$ assigns the term u to the variable x. The test process $[c_1 \wedge \ldots \wedge c_k]^\ell$ terminates if c_i is $\top$ or c_i is $u = v$ where $u =_E v$ for all $i \in \{1, \ldots, k\}$ and otherwise, if some c_i is $u = v$ and $u \neq_E v$, the process deadlocks. The process $\mathsf{in}(x)^\ell$ reads a term u from the public channel and binds it to x and the process $\mathsf{out}(u)^\ell$ outputs a term on the public channel. The processes $R \cdot R'$ sequentially executes R followed by R' whereas the process $R +_p^\ell R'$ behaves like R with probability p and like R' with probability $1 - p$. Note that protocols in our formalism are *simple*; a protocol is said to be simple if there is no principal-level nondeterminism [25].

We will use P and Q to denote processes, which are the parallel composition of a finite set of roles $R_1, \ldots, R_n$, denoted $R_1 \mid \ldots \mid R_n$. For a process Q, $\mathsf{fv}(Q)$ and $\mathsf{bv}(Q)$ denote the set of variables that have some free or bound occurrence in Q, respectively. The formal definition is standard and is omitted for lack of space. Processes containing no free variables are called ground. We restrict our attention to processes that do not contain variables with both free and bound occurrences. That is, for a process Q, $\mathsf{fv}(Q) \cap \mathsf{bv}(Q) = \emptyset$.

Definition 2. *A process $Q = R_1 \mid \ldots \mid R_n$ is said to be well-formed if the following hold.*

1. *Every atomic action and probabilistic choice in Q has a distinct label.*
2. *If label ℓ_1 (resp. ℓ_2) occurs in the role R_i (resp. R_j) for $i, j \in \{1, \ldots, n\}$ then $i \neq j$ iff $[\ell_1] \neq [\ell_2]$.*

For the remainder of this work, processes are assumed to be well-formed. Unless otherwise stated, we will also assume that the labels occurring a role come from the same equivalence class.

Remark 1. For readability, we will omit process labels when they are not relevant in a particular context.

We now present an example illustrating the type of protocols that can be modeled in our process algebra.

Example 2. Using our process syntax, we model a simple threshold mix, as described in Sect. 2.1. We will consider the situation when there two users A_0 and A_1 who want to communicate anonymously through a single mix server M with users B_0 and B_1, respectively. The protocol is built over the equational theory with signature $\mathcal{F}_{\mathsf{aenc}} = \{\mathsf{sk}/1, \mathsf{pk}/1, \mathsf{aenc}/3, \mathsf{adec}/2, \mathsf{pair}/2, \mathsf{fst}/1, \mathsf{snd}/1\}$ and the equations E_{aenc} given below.

$$\mathsf{adec}(\mathsf{aenc}(m, r, \mathsf{pk}(k)), \mathsf{sk}(k)) = m$$
$$\mathsf{fst}(\mathsf{pair}(m_1, m_2)) = m_1$$
$$\mathsf{snd}(\mathsf{pair}(m_1, m_2)) = m_2$$

For generation of their pubic key pairs, the parties A_0, A_1, B_0, B_1 and M will hold private names k_{A_0}, k_{A_1}, k_{B_0}, k_{B_1}, and k_M, respectively. The protocol will also have private names $n_0, n_1, n_2, \ldots$ to model nonces. The nonces n_0 and n_1 are the messages that A_0 and A_1 want to communicate. The behavior of each user and the mix can be described by the roles below (where we use $\langle, \rangle$ in place of pair for succinctness).

$$A_0 = \mathsf{out}(\mathsf{aenc}(\mathsf{aenc}(n_0, n_2, \mathsf{pk}(k_{B_0})), n_4, \mathsf{pk}(k_M)))$$
$$A_1 = \mathsf{out}(\mathsf{aenc}(\mathsf{aenc}(n_1, n_3, \mathsf{pk}(k_{B_1})), n_5, \mathsf{pk}(k_M)))$$
$$M = \mathsf{in}(z_1) \cdot \mathsf{in}(z_2) \cdot$$
$$\mathsf{out}(\langle \mathsf{adec}(z_1, \mathsf{sk}(k_M)), \mathsf{adec}(z_2, \mathsf{sk}(k_M)) \rangle +_{\frac{1}{2}}$$
$$\langle \mathsf{adec}(z_2, \mathsf{sk}(k_M)), \mathsf{adec}(z_1, \mathsf{sk}(k_M)) \rangle)$$

3.3 Partially Observable Markov Decision Processes

POMDPs are used to model processes that exhibit both probabilistic and non-deterministic behavior, where the states of the system are only partially observable. Formally, a POMDP is a tuple $\mathcal{M} = (Z, z_s, \mathsf{Act}, \Delta, \mathcal{O}, \mathsf{obs})$ where Z is a countable set of *states*, $z_s \in Z$ is the *initial state*, Act is a countable set of *actions*, $\Delta : Z \times \mathsf{Act} \hookrightarrow \mathsf{Dist}(Z)$ is a partial function called the *probabilistic transition relation*, $\mathcal{O}$ is a countable set of observations and $\mathsf{obs} : Z \to \mathcal{O}$ is a labeling of states with observations. The POMDP $\mathcal{M}$ is said to be a *fully observable MDP* if obs is an injective function. For a distribution μ over Z, let $\mathsf{support}(\mu) = \{z \in Z \mid \mu(z) > 0\}$. An *execution* ρ of the $\mathcal{M}$ is a finite sequence $z_0 \xrightarrow{\alpha_1} \cdots \xrightarrow{\alpha_m} z_m$ such that $z_0 = z_s$ and for each $i \geq 0$, $z_i \xrightarrow{\alpha_{i+1}} \mu_{i+1}$ and $z_{i+1} \in \mathsf{support}(\mu_{i+1})$.

Such an execution is said to have length m, denoted $|\rho| = m$. The probability an execution ρ in $\mathcal{M}$ is $\mathsf{prob}_{\mathcal{M}}(\rho) = \prod_{i=0}^{|\rho|-1} \Delta(z_i, \alpha_{i+1})(z_i + 1)$ and the set of all executions will be denoted by $\mathsf{Exec}(\mathcal{M})$.

For each state in a POMDP, there is a choice amongst several possible probabilistic transitions. The choice of which probabilistic transition to *trigger* is resolved by an *attacker*. Informally, the process modeled by $\mathcal{M}$ evolves as follows. The process starts in the state z_s. After i execution steps, if the process is in the state z, then the attacker chooses an action α such that $\Delta(z, \alpha) = \mu$ and the process moves to state z' at the $(i+1)$-st step with probability $\mu(z')$. The choice of which action to take is determined by the sequence of observations seen by the attacker.

For an execution $\rho = z_0 \xrightarrow{\alpha_1} \cdots \xrightarrow{\alpha_m} z_m$ we write $\mathsf{tr}(\rho)$ to represent the *trace* of ρ, defined as the sequence $\mathsf{obs}(z_0)\alpha_1 \cdots \alpha_m \mathsf{obs}(z_m)$. The set of all traces is $\mathsf{Trace}(\mathcal{M}) = (\mathcal{O}, \mathsf{Act})^* \cdot \mathcal{O}$ and an attacker is a function $\mathcal{A} : \mathsf{Trace}(\mathcal{M}) \hookrightarrow \mathsf{Act}$. Let $\mathsf{Exec}^{\mathcal{A}}(\mathcal{M}) \subseteq \mathsf{Exec}(\mathcal{M})$ be the smallest set such that $z_s \in \mathsf{Exec}^{\mathcal{A}}(\mathcal{M})$ and if $\rho = \rho' \xrightarrow{\alpha} z \in \mathsf{Exec}^{\mathcal{A}}(\mathcal{M})$ then $\rho' \in \mathsf{Exec}^{\mathcal{A}}(\mathcal{M})$ and $\mathcal{A}(\mathsf{tr}(\rho)) = \alpha$.

State-Based Safety Properties. Given a POMDP $\mathcal{M} = (Z, z_s, \mathsf{Act}, \Delta, \mathcal{O}, \mathsf{obs})$, a set $\Psi \subseteq Z$ is said to be a *state-based safety property*. An execution $\rho = z_0 \xrightarrow{\alpha_1} \cdots \xrightarrow{\alpha_m} z_m$ of $\mathcal{M}$ satisfies Ψ, written $\rho \models \psi$, if $z_j \in \Psi$ for all $0 \le j \le m$. Otherwise $\rho \not\models \psi$. We say that $\mathcal{M}$ satisfies Ψ with probability $\ge p$ against attacker $\mathcal{A}$, denoted $\mathcal{M}^{\mathcal{A}} \models_p \psi$, if the sum of the measures in the set $\{\rho \in \mathsf{Exec}^{\mathcal{A}}(\mathcal{M}) \mid \rho$ is a maximal and $\rho \models \psi\}$ is $\ge p$. $\mathcal{M}$ is said to satisfy Ψ with probability $\ge p$, denoted $\mathcal{M} \models_p \psi$, if for all adversaries $\mathcal{A}$, $\mathcal{M}^{\mathcal{A}} \models_p \psi$.

3.4 Process Semantics

Given a process P, an extended process is a 3-tuple (P, φ, σ) where φ is a frame and σ is a binding substitution. Semantically, a ground process P over equational theory $(\mathcal{F}, E)$ is a POMDP $[\![P]\!] = (Z \cup \{\mathsf{error}\}, z_s, \mathsf{Act}, \Delta, \mathcal{O}, \mathsf{obs})$ where Z is the set of all extended processes $z_s = (P, \emptyset, \emptyset)$, $\mathsf{Act} = (\mathcal{T}(\mathcal{F} \setminus \mathcal{N}, \mathcal{X}_w) \cup \tau) \times \mathcal{L}/\sim$ and $\Delta, \mathcal{O}, \mathsf{obs}$ are defined below. Let $\mu \cdot Q$ denote the distribution μ_1 such that $\mu_1(P', \varphi, \sigma) = \mu(P, \varphi, \sigma)$ if P' is $P \cdot Q$ and 0 otherwise. The distributions $\mu \mid Q$ and $Q \mid \mu$ are defined analogously. For a conjunct c_i $(i \in \{1, \ldots, n\})$ in a test process $[c_1 \wedge \ldots \wedge c_n]$ and a substitution σ we write $c_i \vdash \top$ when c_i is $\top$ or c_i is $u = v$ where $\mathsf{vars}(u, v) \subseteq \mathsf{dom}(\sigma)$ and $u\sigma =_E v\sigma$. We define Δ in Fig. 1, where we write $(P, \varphi, \sigma) \xrightarrow{\alpha} \mu$ if $\Delta((P, \varphi, \sigma), \alpha) = \mu$. For any extended process (P, φ, σ) and action $\alpha \in \mathsf{Act}$, if $\Delta((P, \varphi, \sigma), \alpha)$ is undefined in Fig. 1 then $\Delta((P, \varphi, \sigma), \alpha) = \delta_{\mathsf{error}}$. Note that Δ is well-defined, as roles are deterministic and each equivalence class on labels identifies at most one role. For a frame φ and equational theory E, we write $[\varphi]$ to denote the equivalence class of φ with respect to the static equivalence relation $\equiv_E$. We use EQ to denote the set of all such equivalence classes. Let $\mathcal{O} = \mathsf{EQ}$ and define obs as a function from extended processes to $\mathcal{O}$ such that for any extended process $\eta = (P, \varphi, \sigma)$, $\mathsf{obs}(\eta) = [\varphi]$.

$$\frac{r \in \mathcal{T}(\mathcal{F}\setminus\mathcal{N}, \mathcal{X}_w) \quad \varphi \vdash^r u \quad x \notin \mathsf{dom}(\sigma)}{(\mathsf{in}(x)^\ell, \varphi, \sigma) \xrightarrow{(r,[\ell])} \delta_{(0,\varphi,\sigma\cup\{x\mapsto u\})}} \text{ IN}$$

$$\frac{\mathsf{vars}(u) \subseteq \mathsf{dom}(\sigma) \quad i = |\mathsf{dom}(\varphi)| + 1}{(\mathsf{out}(u)^\ell, \varphi, \sigma) \xrightarrow{(\tau,[\ell])} \delta_{(0,\varphi\cup\{w_{(i,[\ell])}\mapsto u\sigma\},\sigma)}} \text{ OUT}$$

$$\frac{\forall i \in \{1,\ldots,n\}, c_i \vdash \top}{([c_1 \wedge \ldots \wedge c_n]^\ell, \varphi, \sigma) \xrightarrow{(\tau,[\ell])} \delta_{(0,\varphi,\sigma)}} \text{ TEST}$$

$$\frac{\mathsf{vars}(u) \subseteq \mathsf{dom}(\sigma) \quad x \notin \mathsf{dom}(\sigma)}{((x := u)^\ell, \varphi, \sigma) \xrightarrow{(\tau,[\ell])} \delta_{(0,\varphi,\sigma\cup\{x\mapsto u\sigma\})}} \text{ ASGN}$$

$$\frac{}{(Q_1 +_p^\ell Q_2, \varphi, \sigma) \xrightarrow{(\tau,[\ell])} \delta_{(Q_1,\varphi,\sigma)} +_p \delta_{(Q_2,\varphi,\sigma)}} \text{ PROB}$$

$$\frac{x \notin \mathsf{dom}(\sigma) \quad n \text{ is a fresh name}}{(\nu x^\ell, \varphi, \sigma) \xrightarrow{(\tau,[\ell])} \delta_{(0,\varphi,\sigma\cup\{x\mapsto n\})}} \text{ NEW}$$

$$\frac{Q_0 \neq 0 \quad (Q_0,\varphi,\sigma) \xrightarrow{\alpha} \mu}{(Q_0 \cdot Q_1, \varphi, \sigma) \xrightarrow{\alpha} \mu \cdot Q_1} \text{ SEQ}$$

$$\frac{(Q_0,\varphi,\sigma) \xrightarrow{\alpha} \mu}{(0 \cdot Q_0, \varphi, \sigma) \xrightarrow{\alpha} \mu} \text{ NULL}$$

$$\frac{(Q_0,\varphi,\sigma) \xrightarrow{\alpha} \mu}{(Q_0 \mid Q_1, \varphi, \sigma) \xrightarrow{\alpha} \mu \mid Q_1} \text{ PAR}_\mathrm{L}$$

$$\frac{(Q_1,\varphi,\sigma) \xrightarrow{\alpha} \mu}{(Q_0 \mid Q_1, \varphi, \sigma) \xrightarrow{\alpha} Q_0 \mid \mu} \text{ PAR}_\mathrm{R}$$

Fig. 1. Process semantics.

Definition 3. *An extended process (P, φ, σ) preserves the secrecy of a term u in the equational theory $(\mathcal{F}, E)$, denoted $(P, \varphi, \sigma) \models_E u$, if there is no $r \in \mathcal{T}(\mathcal{F}\setminus\mathcal{N}, \mathsf{dom}(\varphi))$ such that $\varphi \vdash_E^r u\sigma$. We write $\mathsf{secret}(u)$, to represent the set of states of $[\![P]\!]$ that preserve the secrecy of u and $\mathsf{secret}(\{u_1,\ldots,u_n\})$ to denote $\mathsf{secret}(u_1) \cap \ldots \cap \mathsf{secret}(u_n)$.*

Remark 2. For a process P and terms $u_1,\ldots,u_n$, $\mathsf{secret}(\{u_1,\ldots,u_n\})$ is a state-based safety property of $[\![P]\!]$. For a probability p, we will write $P \models_{E,p} \mathsf{secret}(u_1,\ldots,u_n)$, if $[\![P]\!] \models_p \mathsf{secret}(\{u_1,\ldots,u_n\})$.

Example 3. Consider the mix-net protocol $P = A_0 \mid A_1 \mid M$ defined in Example 2. The protocol is designed to ensure that the messages output by the mix cannot be linked to the original sends with high probability. That is, the adversary should be able to do no better than "guess" which output message belongs to which sender. This hypothesis is violated if, for an output of the mix, the adversary can identify the sender of the message with probability $> \frac{1}{2}$. We can model this property in our framework by adding, for each $i \in \{0,1\}$, a role

$$S_i = \mathsf{in}(z_i') \cdot [z_i' = \mathsf{aenc}(n_i, n_{i+2}, \mathsf{pk}(k_{B_i}))] \cdot \mathsf{out}(s_i)$$

to the process, where s_i is a private name. The protocol P preserves the anonymity of sender A_i if $(A_0 \mid A_1 \mid M \mid S_0 \mid S_1) \models_{E_\mathsf{aenc}, \frac{1}{2}} \mathsf{secret}(s_i)$.

4   Model Checking Algorithm

As seen in Sect. 3, analyzing randomized protocols requires reasoning about their underlying semantic objects, POMDPs. In particular, we are interested in finding an attacker for a given POMDP that maximizes the probability of reaching a set of target (bad) states. Unfortunately, techniques for solving reachability problems in POMDPs are far less efficient than those for Markov Decision Processes (MDPs), the fully observable counterpart to POMDPs (where attackers are a function from *executions* to actions). The reason for the added complexity is that at any given point in the execution of a POMDP, the attacker only knows a distribution over the current state. Further, an attacker for a POMDP needs to define a consistent strategy across all executions that produce the same sequence of observations. The actions chosen in one branch of an execution may affect the actions that can be made in another branch of the same execution. By contrast, when trying to maximize a reachability probability in an MDP, one can make a local decision about which action maximizes the probability of reaching the target states.

Several results [18, 26] corroborate this story, showing that many key verification problems for POMDPs are undecidable. Although various solution techniques have been proposed [12], and there have been successful applications to AI and planning [14], tractable reasoning about POMDPs typically relies on approximation techniques or simplifications to the model (discounts). Complicating matters further, randomized security protocols induce POMDPs that are infinitely branching. At every transition corresponding to protocol input, an infinite number of possible recipes can be supplied by a Dolev−Yao attacker. Taming the state space explosion that results from this infinite branching on inputs is a huge challenge, even in the non-randomized case. We adopt the philosophy of the SATMC [5] tool, in that, we will search for bounded attacks. That is, our tool answers the question; for a given input recipe depth k, what is the maximum probability of reaching a set of target states? The assumption of bounded recipe depth allows randomized security protocol to be modeled by POMDPs that are finite branching.

One of the most successful techniques in the approximation of optimal attackers for POMDPs is to translate a POMDP $\mathcal{M}$ into a fully observable belief MDP $\mathcal{B}(\mathcal{M})$ that emulates it. One can then analyze $\mathcal{B}(\mathcal{M})$ to infer properties of $\mathcal{M}$. The states of $\mathcal{B}(\mathcal{M})$ are probability distributions over the states of $\mathcal{M}$. Further, given a state b of $\mathcal{B}(\mathcal{M})$, if states z_1, z_2 of $\mathcal{M}$ are such that $b(z_1), b(z_2)$ are non-zero then z_1 and z_2 must have the same observation. Hence, by abuse of notation, we can define $\mathsf{obs}(b)$ to be $\mathsf{obs}(z)$ if $b(z) \neq 0$. Intuitively, an execution $\rho = b_0 \xrightarrow{\alpha_1} b_1 \xrightarrow{\alpha_2} \cdots \xrightarrow{\alpha_m} b_m$ of $\mathcal{B}(\mathcal{M})$ corresponds to the set of all executions ρ' of $\mathcal{M}$ such that $\mathsf{tr}(\rho') = \mathsf{obs}(b_0)\alpha_1\mathsf{obs}(b_1)\alpha_2 \cdots \alpha_m\mathsf{obs}(b_m)$. The measure of execution ρ in $\mathcal{B}(\mathcal{M})$ is exactly $\mathsf{prob}_{\mathcal{M}}(\mathsf{obs}(b_0)\alpha_1\mathsf{obs}(b_1)\alpha_2 \cdots \alpha_m\mathsf{obs}(b_m))$.

The initial state of $\mathcal{B}(\mathcal{M})$ is the distribution that assigns 1 to the initial state of $\mathcal{M}$. Intuitively, on a given state b of $\mathcal{B}(\mathcal{M})$ and an action α, there is at most one successor state $b^{\alpha,o}$ for each observation o. The probability of transitioning from b to $b^{\alpha,o}$ is the probability that o is observed given that the distribution on the states of $\mathcal{M}$ is b and action α is performed; $b^{\alpha,o}(z)$ is the conditional probability that the actual state of the POMDP is z. The formal definition follows.

Definition 4. *Let* $\mathcal{M} = (Z, z_s, \mathsf{Act}, \Delta, \mathcal{O}, \mathsf{obs})$ *be a POMDP. The belief MDP of* $\mathcal{M}$, *denoted* $\mathcal{B}(\mathcal{M})$, *is the tuple* $(\mathsf{Dist}(Z), \delta_{z_s}, \mathsf{Act}, \Delta^{\mathcal{B}})$ *where* $\Delta^{\mathcal{B}}$ *is defined as follows. For* $b \in \mathsf{Dist}(Z)$, *action* $\alpha \in \mathsf{Act}$ *and* $o \in \mathcal{O}$, *let*

$$p_{b,\alpha,o} = \sum_{z \in Z} b(z) \cdot \left(\sum_{z' \in Z \wedge \mathsf{obs}(z')=o} \Delta(z,\alpha)(z') \right).$$

$\Delta^{\mathcal{B}}(b, \alpha)$ *is the unique distribution such that for each* $o \in \mathcal{O}$, *if* $p_{b,\alpha,o} \neq 0$ *then* $\Delta^{\mathcal{B}}(b, \alpha)(b^{\alpha,o}) = p_{b,\alpha,o}$ *where for all* $z' \in Z$,

$$b^{\alpha,o}(z') = \begin{cases} \frac{\sum_{z \in Z} b(z) \cdot \Delta(z,\alpha)(z')}{p_{b,\alpha,o}} & \text{if } \mathsf{obs}(z') = o \\ 0 & \text{otherwise} \end{cases}.$$

This definition results in a correspondence between the maximal reachability probabilities in a POMDP $\mathcal{M}$ and the belief MDP $\mathcal{B}(\mathcal{M})$ it induces. The following proposition, due to Norman et al. [40], makes this correspondence precise. In the result below, for a POMDP (resp. MDP) $\mathcal{M}$ and a set of observations O (resp. states T), we write $\mathsf{prob}_{\mathcal{M}}^{max}(O)$ (resp. $\mathsf{prob}_{\mathcal{M}}^{max}(T)$) to denote the maximum probability with which $\mathcal{M}^A$ reaches states with observations in O (resp. states from T) for any adversary $\mathcal{A}$.

Proposition 1. *Let* $\mathcal{M} = (Z, z_s, \mathsf{Act}, \Delta, \mathcal{O}, \mathsf{obs})$ *be a POMDP,* $O \subseteq \mathcal{O}$ *and* $T_O = \{b \in \mathsf{Dist}(Z) \mid \forall z \in Z.(b(z) > 0 \Rightarrow \mathsf{obs}(z) \in O)\}$. *Then* $\mathsf{prob}_{\mathcal{M}}^{max}(O) = \mathsf{prob}_{\mathcal{B}(\mathcal{M})}^{max}(T_O)$.

In general, belief MDPs are defined over a continuous state space; even simple POMDP models can yield an infinite number of distributions on states. It is this continuous state space that makes belief MDPs difficult to analyze. Fortunately, the calculus from Sect. 3.2 doesn't include an operator for replication. This means that protocol executions are of a fixed length and can be encoded as acyclic POMDPs that reach a set of finite absorbing states after a bounded number of actions. However, even for acyclic POMDPs, the number of reachable belief states can grow much larger than the number of states in the original POMDP.

Let Q be a randomized security protocol such that $[\![Q]\!] = (Z, z_s, \mathsf{Act}, \Delta, \mathcal{O}, \mathsf{obs})$. Define $[\![Q_d]\!] = (Z, z_s, \mathsf{Act}_d, \Delta_d, \mathcal{O}, \mathsf{obs})$ where every $\alpha \in \mathsf{Act}_d$ is such that $\mathsf{depth}(\alpha) \leq d$ and for all $z \in Z$, $\Delta_d(z, \alpha) = \Delta(z, \alpha)$ if $\alpha \in \mathsf{Act}_d$ and otherwise $\Delta_d(z, \alpha)$ is undefined. For a security protocol Q, probability p and safety property ψ, the *bounded model checking problem* for depth d is to determine if $[\![Q_d]\!] \models_p \psi$. As described above, $[\![Q_d]\!]$ can be translated into a finite acyclic fully observable belief MDP $\mathcal{B}([\![Q_d]\!])$. By analyzing $\mathcal{B}([\![Q_d]\!])$, one can generate an attacker for $[\![Q_d]\!]$ that optimizes the probability of reaching a target set of states $Z \setminus \psi$. These optimal reachability probabilities can be computed using Algorithm 1, where we assume a finite set of absorbing states B_{abs}. The algorithm works by recursively computing the maximum probability of attack by exploring states in a depth-first fashion. Such an approach can avoid exploring many redundant portions of the state space.

The correctness of our algorithm, which follows from Proposition 1, is given below.

Algorithm 1. *On-the-fly model checking of safety properties in finite-length belief MDPs.*

1: **procedure** MAXATTACK(beliefState b, targetStates T)
2:      $p \leftarrow 0$
3:      **if** $b \in B_{\mathsf{abs}}$ **then**
4:          **for** $, \in \mathsf{support}(b)$ **do**
5:              **if** $\in T$ **then**
6:                  $p \leftarrow p + b()$
7:          **return** p
8:      **for** $\alpha \in \mathsf{Act}$ **do**
9:          **for** $o \in \mathcal{O}$ **do**
10:             $p \leftarrow \max(p, \text{MAXATTACK}(b^{\alpha,o}, T))$
11:             **if** $p == 1$ **then**
12:                 **return** 1
13:     **return** p

Theorem 1. *Let Q be a protocol and $d \in \mathbb{N}$ be such that $\llbracket Q_d \rrbracket = (Z, z_s, \mathsf{Act}_d, \Delta, \mathcal{O}, \mathsf{obs})$. For a given probability p and state-based safety property $\psi \subseteq Z$, if $\llbracket Q_d \rrbracket \models_p \psi$ iff MAXATTACK$(\delta_z, Z \setminus \psi) \leq 1 - p$ for the belief MDP $\mathcal{B}(\llbracket Q_d \rrbracket)$.*

5   Tool Description and Evaluation

The algorithm for checking safety in randomized security protocols is implemented in the tool, SPAN. We refer the reader to [7] for a detailed discussion of the implementation and evaluation of SPAN. We describe the salient features briefly.

Implementation. As described in Sect. 4, the fundamental routine of SPAN translates a randomized security protocol into a belief MDP. Each translation step requires operations from term rewriting as well as solving the static equivalence and deduction problems on protocol frames. Currently, SPAN supports two external engines for solving the static equivalence and deduction questions: KISS [3] and AKISS [15]. KISS tool supports sub-term convergent theories, while the AKISS tool supports more general optimally reducing theories and the AC operation XOR. SPAN implements its own unification algorithm for convergent equational theories for its term-rewriting engine. For rewriting in the presence of AC operations, support for integration with Maude [24,30] is also included. Because attacks on randomized protocols are trees (as opposed to sequences) attacks are exported to DOT format, which can be rendered visually using the graphviz framework [1].

Evaluation. We evaluated SPAN on a variety of protocols. Our experiments were conducted on an Intel core i7 dual quad-core processor at 2.67 GHz with 12 GB of RAM. The host operating system was 64 bit Ubuntu 16.04.3 LTS. The examples

Table 1. *Experimental results for safety properties.* Columns 1–5 describe the example under test, where column 2 is the number of users in the protocol, column 3 is maximum recipe depth, column 4 is the maximum attack probability and column 5 is the security threshold: if the value of column 4 exceeds the value of column 5, then an attack was found. Columns 6 and 7 give the running times (in seconds) under the KISS and Akiss, respectively. Column 8 reports the number of belief states explored during the model checking procedure. All test were conducted using Maude 2.7.1 as the term rewriting engine. For protocols with requiring equational theories with XOR we write n/s (not supported) for the KISS engine.

1	2	3	4	5	6	7	8
PROTOCOL	PARTIES	DEPTH	ATTACK	THRESHOLD	TIME (S)		BELIEFS
					w/ KISS	w/ AKISS	
DC-net	2	10	1/2	1/2	n/s	23	110
Threshold Mix	4	10	1	1/4	22	70	49
Cascade Mix	2	5	1	1/2	917	2832	55303
Pool Mix	3	5	2/3	1/3	1824	6639	26273
FOO 92 (corrected)	2	10	3/4	3/4	321	918	1813
Prêt à Voter	2	10	7/8	3/4	n/s	288	103

that we verified were sender anonymity in Dining Cryptographers-Net [20,35], threshold mixes [22] and pool mixes [46–48], and vote privacy in FOO voting protocol [32] and Prêt à Voter protocol [42]. We attempted to verify all protocols with a recipe depth of 10; however, for some examples, SPAN did not terminate within a reasonable time-bound. In such cases, we report the time for a recipe depth of 5. Our experimental results are summarized in Table 1. As mentioned above, mixes are vulnerable to active attacks, and our tool was able to capture these attacks. For the FOO voting protocol, we implemented the anonymous channels using threshold mixes. In the previous automated analysis of FOO voting protocol (See [15], for example), perfectly anonymous channels are assumed to exist. This abstraction misses possible attacks. For example, if a threshold mix is used to implement the FOO protocol, then SPAN found an attack on vote privacy that exploits the flooding attack on mixes. A similar attack has also been previously reported in [6], which carries out the analysis of FOO voting protocol in the computational model, and was discovered by hand. We propose corrections to the FOO protocol to avoid such attacks. Finally, in order to capture the attack on Prêt à Voter protocol described above, we assumed that the sum of two hashes is even with probability $\frac{3}{4}$ and odd with probability $\frac{1}{4}$.

6   Conclusion

We present a bounded model checking algorithm to verify safety properties of acyclic randomized security protocols. As randomized security protocols are naturally modeled as POMDPs, we adapt the belief MDP construction from POMPDP literature in the design of the algorithm. The algorithm exploits the acyclic nature of the protocols considered and constructs the belief MDP by traversing the belief MDP in a Depth First Search fashion. The algorithm is implemented as an extension of SPAN. Our experiments demonstrate the effectiveness of the tool in uncovering previously unknown attacks in protocols.

We plan to investigate the use of partial order reduction and symmetry reduction techniques to combat the state explosion problem. We also plan to investigate the verification of randomized security protocols without any restriction of recipe sizes. Another line of investigation that we plan to pursue is the verification of cyclic randomized security protocols.

Acknowledgements. Andre Scedrov's foundational work on formal analysis of security protocols has been an unmistakable inspiration for us, and we thank him for his mentorship. Rohit Chadha thanks Andre Scedrov for introducing him to the exciting and challenging field of security protocol analysis, and his invaluable counsel.

Rohit Chadha was partially supported by grants NSF 1553548 CNS and NSF CCF 1900924. Mahesh Viswanathan was partially supported by NSF CCF 1901069.

References

1. Graphviz. https://www.graphviz.org/
2. Abadi, M., Fournet, C.: Mobile values, new names, and secure communication. In: ACM SIGPLAN Notices, vol. 36, pp. 104–115. ACM (2001)
3. Abadi, M., Cortier, V.: Deciding knowledge in security protocols under equational theories. Theor. Comput. Sci. **367**(1), 2–32 (2006)
4. Adida, B.: Helios: web-based open-audit voting. In: USENIX Security Symposium, vol. 17, pp. 335–348 (2008)
5. Armando, A., Compagna, L.: SAT-based model-checking for security protocols analysis. Int. J. Inf. Secur. **7**(1), 3–32 (2008)
6. Bana, G., Chadha, R., Eeralla, A.K.: Formal analysis of vote privacy using computationally complete symbolic attacker. In: Lopez, J., Zhou, J., Soriano, M. (eds.) ESORICS 2018. LNCS, vol. 11099, pp. 350–372. Springer, Cham (2018). https://doi.org/10.1007/978-3-319-98989-1_18
7. Bauer, M.S.: Analysis of randomized security protocols. Ph.D. thesis, University of Illinois at Urbana-Champaign (2018)
8. Bauer, M.S., Chadha, R., Prasad Sistla, A., Viswanathan, M.: Model checking indistinguishability of randomized security protocols. In: Chockler, H., Weissenbacher, G. (eds.) CAV 2018. LNCS, vol. 10982, pp. 117–135. Springer, Cham (2018). https://doi.org/10.1007/978-3-319-96142-2_10
9. Bauer, M.S., Chadha, R., Viswanathan, M.: Composing protocols with randomized actions. In: Piessens, F., Viganò, L. (eds.) POST 2016. LNCS, vol. 9635, pp. 189–210. Springer, Heidelberg (2016). https://doi.org/10.1007/978-3-662-49635-0_10
10. Ben-Or, M., Goldreich, O., Micali, S., Rivest, R.L.: A fair protocol for signing contracts. IEEE Trans. Inf. Theory **36**(1), 40–46 (1990)
11. Blanchet, B., Abadi, M., Fournet, C.: Automated verification of selected equivalences for security protocols. J. Log. Algebr. Program. **75**(1), 3–51 (2008)
12. Braziunas, D.: POMDP Solution Methods. University of Toronto (2003)
13. Canetti, R., et al.: Task-structured probabilistic I/O automata. In: Discrete Event Systems (2006)

14. Cassandra, A.R.: A survey of POMDP applications. In: Working notes of AAAI 1998 fall Symposium on Planning with Partially Observable Markov Decision Processes, vol. 1724 (1998)
15. Chadha, R., Cheval, V., Ciobâcă, Ş., Kremer, S.: Automated verification of equivalence properties of cryptographic protocol. ACM Trans. Comput. Log. 17(4), 1–32 (2016)
16. Chadha, R., Sistla, A.P., Viswanathan, M.: Model checking concurrent programs with nondeterminism and randomization. In: Foundations of Software Technology and Theoretical Computer Science, pp. 364–375 (2010)
17. Chadha, R., Sistla, A.P., Viswanathan, M.: Verification of randomized security protocols. In: Logic in Computer Science, pp. 1–12. IEEE (2017)
18. Chatterjee, K., Chmelík, M., Tracol, M.: What is decidable about partially observable Markov decision processes with omega-regular objectives. J. Comput. Syst. Sci. 82(5), 878–911 (2016)
19. Chatzikokolakis, K., Palamidessi, C.: Making random choices invisible to the scheduler. Information and Computation (2010, to appear)
20. Chaum, D.: The dining cryptographers problem: unconditional sender and recipient untraceability. J. Cryptol. 1(1), 65–75 (1988)
21. Chaum, D., Ryan, P.Y.A., Schneider, S.: A practical voter-verifiable election scheme. In: di Vimercati, S.C., Syverson, P., Gollmann, D. (eds.) ESORICS 2005. LNCS, vol. 3679, pp. 118–139. Springer, Heidelberg (2005). https://doi.org/10.1007/11555827_8
22. Chaum, D.L.: Untraceable electronic mail, return addresses, and digital pseudonyms. Commun. ACM 24(2), 84–90 (1981)
23. Cheung, L.: Reconciling nondeterministic and probabilistic choices. Ph.D. thesis, Radboud University of Nijmegen (2006)
24. Clavel, M., et al.: Maude: Specification and programming in rewriting logic. Theor. Comput. Sci. 285(2), 187–243 (2002)
25. Cortier, V., Delaune, S.: A method for proving observational equivalence. In: Computer Security Foundations, pp. 266–276 (2009)
26. de Alfaro, L.: The verification of probabilistic systems under memoryless partial-information policies is hard. Technical report (1999)
27. Dehnert, C., Junges, S., Katoen, J.P., Volk, M.: A storm is coming: a modern probabilistic model checker. In: Majumdar, R., Kunčak, V. (eds.) Computer Aided Verification CAV 2017. LNCS, vol. 10427, pp. 592-600. Springer, Cham (2017). https://doi.org/10.1007/978-3-319-63390-9_31
28. Delaune, S., Kremer, S., Ryan, M.: Verifying privacy-type properties of electronic voting protocols. J. Comput. Secur. 17(4), 435–487 (2009)
29. Dingledine, R., Mathewson, N., Syverson, P.: Tor: the second-generation onion router. Technical report, DTIC Document (2004)
30. Escobar, S., Meadows, C., Meseguer, J.: Maude-NPA: cryptographic protocol analysis modulo equational properties. In: Aldini, A., Barthe, G., Gorrieri, R. (eds.) FOSAD 2007-2009. LNCS, vol. 5705, pp. 1–50. Springer, Heidelberg (2009). https://doi.org/10.1007/978-3-642-03829-7_1
31. Even, S., Goldreich, O., Lempel, A.: A randomized protocol for signing contracts. Commun. ACM 28(6), 637–647 (1985)
32. Fujioka, A., Okamoto, T., Ohta, K.: A practical secret voting scheme for large scale elections. In: Seberry, J., Zheng, Y. (eds.) AUSCRYPT 1992. LNCS, vol. 718, pp. 244–251. Springer, Heidelberg (1993). https://doi.org/10.1007/3-540-57220-1_66
33. Garcia, F.D., Van Rossum, P., Sokolova, A.: Probabilistic anonymity and admissible schedulers. arXiv preprint arXiv:0706.1019 (2007)

34. Goldschlag, D.M., Reed, M.G., Syverson, P.F.: Hiding routing information. In: Workshop on Information Hiding, pp. 137–150 (1996)
35. Golle, P., Juels, A.: Dining cryptographers revisited. In: Cachin, C., Camenisch, J.L. (eds.) EUROCRYPT 2004. LNCS, vol. 3027, pp. 456–473. Springer, Heidelberg (2004). https://doi.org/10.1007/978-3-540-24676-3_27
36. Goubault-Larrecq, J., Palamidessi, C., Troina, A.: A probabilistic applied pi-calculus. In: Shao, Z. (ed.) APLAS 2007. LNCS, vol. 4807, pp. 175–190. Springer, Heidelberg (2007). https://doi.org/10.1007/978-3-540-76637-7_12
37. Gunter, C.A., Khanna, S., Tan, K., Venkatesh, S.S.: DoS protection for reliably authenticated broadcast. In: Network and Distributed System Security (2004)
38. Kremer, S., Ryan, M.: Analysis of an electronic voting protocol in the applied pi calculus. In: Sagiv, M. (ed.) ESOP 2005. LNCS, vol. 3444, pp. 186–200. Springer, Heidelberg (2005). https://doi.org/10.1007/978-3-540-31987-0_14
39. Kwiatkowska, M., Norman, G., Parker, D.: PRISM 4.0: verification of probabilistic real-time systems. In: Gopalakrishnan, G., Qadeer, S. (eds.) CAV 2011. LNCS, vol. 6806, pp. 585–591. Springer, Heidelberg (2011). https://doi.org/10.1007/978-3-642-22110-1_47
40. Norman, G., Parker, D., Zou, X.: Verification and control of partially observable probabilistic systems. Real-Time Syst. 53(3), 354–402 (2017). https://doi.org/10.1007/s11241-017-9269-4
41. Reiter, M.K., Rubin, A.D.: Crowds: anonymity for web transactions. ACM Trans. Inf. Syst. Secur. 1(1), 66–92 (1998)
42. Ryan, P.Y.A., Bismark, D., Heather, J., Schneider, S., Xia, Z.: Prêt à voter: a voter-verifiable voting system. IEEE Trans. Inf. Forensics Secur. 4(4), 662–673 (2009)
43. Bauer, M.S., Chadha, R., Viswanathan, M.: Modular verification of protocol equivalence in the presence of randomness. In: Foley, S.N., Gollmann, D., Snekkenes, E. (eds.) ESORICS 2017. LNCS, vol. 10492, pp. 187–205. Springer, Cham (2017). https://doi.org/10.1007/978-3-319-66402-6_12
44. Santin, A.O., Costa, R.G., Maziero, C.A.: A three-ballot-based secure electronic voting system. Secur. Priv. 6(3), 14–21 (2008)
45. Schmidt, B., Meier, S., Cremers, C., Basin, D.: Automated analysis of Diffie-Hellman protocols and advanced security properties. In: Computer Security Foundations, pp. 78–94 (2012)
46. Serjantov, A., Dingledine, R., Syverson, P.: From a trickle to a flood: active attacks on several mix types. In: Petitcolas, F.A.P. (ed.) IH 2002. LNCS, vol. 2578, pp. 36–52. Springer, Heidelberg (2003). https://doi.org/10.1007/3-540-36415-3_3
47. Serjantov, A., Newman, R.E.: On the anonymity of timed pool mixes. In: Gritzalis, D., De Capitani di Vimercati, S., Samarati, P., Katsikas, S. (eds.) SEC 2003. ITIFIP, vol. 122, pp. 427–434. Springer, Boston, MA (2003). https://doi.org/10.1007/978-0-387-35691-4_41
48. Serjantov, A., Sewell, P.: Passive attack analysis for connection-based anonymity systems. In: Snekkenes, E., Gollmann, D. (eds.) ESORICS 2003. LNCS, vol. 2808, pp. 116–131. Springer, Heidelberg (2003). https://doi.org/10.1007/978-3-540-39650-5_7
49. Shmatikov, V.: Probabilistic analysis of anonymity. In: Computer Security Foundations, pp. 119–128. IEEE (2002)

Logic and Language

Andre Scedrov

Glyn Morrill[✉]

Universitat Politècnica de Catalunya, Barcelona, Spain
morrill@cs.upc.edu

Amongst the subjects I found most difficult at school were languages (French and Latin). Like a moth to the flame my research has been dedicated to grammar: formal grammar, computational grammar, logical grammar, and mathematical grammar. As a youth I sought literal translation, and my research has reflected this in assuming interlingual semantics such as is provided by Montague-like higher-order intensional logic. But my experience of living three decades in the bilingual (Catalan and Spanish) community of Barcelona is that in the end each language is a world unto itself and that translation in general is a holistic art rather than a reductionist science.

When I began as a postgraduate in Edinburgh in 1984–85 Mark Steedman was circulating a draft of a paper later published in *Natural Language and Linguistic Theory* in 1987 under the title 'Combinatory grammars and parasitic gaps'. Parasitic gaps, such as that in (1c):

(1) a. the paper that$_i$ John filed e_i without reading the guidelines

    b. ?the guidelines that John filed the paper without reading e_i

    c. the paper that$_i$ John filed e_i without reading e_i

exhibit finely controlled syntactic-semantic contraction. This counterexample to linearity been a constant source of fascination to me.

I came to the Polytecnic University of Catalunya in 1991 and here I have sought to practice linguistics as an exact, mathematicised, science. For much of this time I have had the benefit and privaledge to work with the mathematician and linguist Oriol Valentín. Initially unbeknownst to me, in these last five years a powerhouse team comprising Max Kanovich, Stepan Kuznetsov and Andre Scedrov began working on the subtle formal properties of controlled contraction in logical grammar.

At the 2016 European Summer School in Logic, Language and Information in Bolzano, Stepan invited me on behalf of Andre to visit the Math department at the University of Pennsylvania. This I did in February 2017, and there I met Andre. Kind, relaxed, efficient, fair, intelligent, pleasant, amusing, and interesting. And also devout. Several times he invited me into his home to dine with Max, Stepan, and his family.

During that February visit the four of us wrote an article on the polynomial decidability of the Lambek calculus with bracket modalities (for Formal Grammar 2017) generalising the polynomial algorithm for the Lambek Calculus with product of Pentus; and Stepan joined Oriol and I in completing an article on count invariance (for Mathematics of Language 2017) including infinitary counts for (sub)exponentials. In a subsequent paper Max, Stepan, Andre and I

© Springer Nature Switzerland AG 2020
V. Nigam et al. (Eds.): Scedrov Festschrift, LNCS 12300, pp. 187–188, 2020.
https://doi.org/10.1007/978-3-030-62077-6_13

investigated bracket induction for the Lambek calculus with bracket modalities (Formal Grammar 2018), whereby brackets need not be included in the parsing input but are discovered (induced) in the course of processing.

In recent work Max, Stepan and Andre have resumed studies on undecidability of controlled contraction, showing that both of my recent formulations are undecidable. I think that typically in mathematical linguistics an undecidability or high complexity result invites questioning whether the formal source of the complexity as revealed by the proof is linguistically motivated and, if not, how this complexity of the formalism can be lowered while maintaining linguistic adequacy. In this respect Max, Stepan and Andre suggest that the set of contractable formulas might be limited to atoms and I think this would be a good direction.

Gender Bias in Neural Natural Language Processing

Kaiji Lu[1]([✉]), Piotr Mardziel[1], Fangjing Wu[1,2], Preetam Amancharla[1,3], and Anupam Datta[1]

[1] Carnegie Mellon University, Moffiet Field, Pittsburgh, CA, USA
kaijil@andrew.cmu.edu
[2] Facebook, Menlo Park, CA, USA
[3] The Yes Platform, Burlingame, CA, USA

Abstract. We examine whether neural natural language processing (NLP) systems reflect historical biases in training data. We define a general benchmark to quantify gender bias in a variety of neural NLP tasks. Our empirical evaluation with state-of-the-art neural coreference resolution and textbook RNN-based language models trained on benchmark data sets finds significant gender bias in how models view occupations. We then mitigate bias with *counterfactual data augmentation (CDA)*: a generic methodology for corpus augmentation via causal interventions that breaks associations between gendered and gender-neutral words. We empirically show that CDA effectively decreases gender bias while preserving accuracy. We also explore the space of mitigation strategies with CDA, a prior approach to word embedding debiasing (WED), and their compositions. We show that CDA outperforms WED, drastically so when word embeddings are trained. For pre-trained embeddings, the two methods can be effectively composed. We also find that as training proceeds on the original data set with gradient descent the gender bias grows as the loss reduces, indicating that the optimization encourages bias; CDA mitigates this behavior.

Keywords: Machine learning · Deep learning · Fairness · Natural language processing

1 Introduction

Natural language processing (NLP) with neural networks has grown in importance over the last few years. They provide state-of-the-art models for tasks like coreference resolution, language modeling, and machine translation [4,5,10,11,14] However, since these models are trained on human language texts, a natural question is whether they exhibit bias based on gender or other characteristics, and, if so, how should this bias be mitigated. This is the question that we address in this paper.

F. Wu and P. Amancharla—Work done while at Carnegie Mellon University.

© Springer Nature Switzerland AG 2020
V. Nigam et al. (Eds.): Scedrov Festschrift, LNCS 12300, pp. 189–202, 2020.
https://doi.org/10.1007/978-3-030-62077-6_14

Prior work provides evidence of bias in autocomplete suggestions [13] and differences in accuracy of speech recognition based on gender and dialect [23] on popular online platforms. Word embeddings, initial pre-processors in many NLP tasks, embed words of a natural language into a vector space of limited dimension to use as their semantic representation. [2] and [3] observed that popular word embeddings including *word2vec* [19] exhibit gender bias mirroring stereotypical gender associations such as the eponymous [2] "Man is to computer programmer as Woman is to homemaker".

Yet the question of how to measure bias in a general way for neural NLP tasks has not been studied. Our first contribution is a general benchmark to quantify gender bias in a variety of neural NLP tasks. Our definition of bias loosely follows the idea of causal testing: matched pairs of individuals (instances) that differ in only a targeted concept (like gender) are evaluated by a model and the difference in outcomes (or scores) is interpreted as the causal influence of the concept in the scrutinized model. The definition is parametric in the scoring function and the target concept. Natural scoring functions exist for a number of neural natural language processing tasks.

We instantiate the definition for two important tasks—coreference resolution and language modeling. Coreference resolution is the task of finding words and expressions referring to the same entity in a natural language text. The goal of language modeling is to model the distribution of word sequences. For neural coreference resolution models, we measure the gender coreference score disparity between gender-neutral words and gendered words like the disparity between "doctor" and "he" relative to "doctor" and "she" pictured as edge weights in Fig. 1a. For language models, we measure the disparities of emission log-likelihood of gender-neutral words conditioned on gendered sentence prefixes as is shown in Fig. 1b. Our empirical evaluation with state-of-the-art neural coreference resolution and textbook RNN-based language models [4,14,25] trained on benchmark datasets finds gender bias in these models[1].

Next we turn our attention to mitigating the bias. [2] introduced a technique for *debiasing* word embeddings which has been shown to mitigate unwanted associations in analogy tasks while preserving the embedding's semantic properties. Given their widespread use, a natural question is whether this technique is sufficient to eliminate bias from downstream tasks like coreference resolution and language modeling. As our second contribution, we explore this question empirically. We find that while the technique does reduce bias, the residual bias is considerable. We further discover that debiasing models that make use of embeddings that are co-trained with their other parameters [4,25] exhibit a significant drop in accuracy.

Our third contribution is *counterfactual data augmentation (CDA)*: a generic methodology to mitigate bias in neural NLP tasks. For each training instance,

---

[1] Note that these results have practical significance. Both coreference resolution and language modeling are core natural language processing tasks in that they form the basis of many practical systems for information extraction [28], text generation [8], speech recognition [9] and machine translation [1].

$$\overbrace{\hspace{3cm}}^{5.08}$$
$1_\square$: The **doctor** ran because **he** is late.
$$\underbrace{\hspace{2.5cm}}_{1.99}$$

$1_\bigcirc$: The **doctor** ran because **she** is late.
$$\overline{\hspace{2cm}}^{-0.44}$$

$2_\square$: The **nurse** ran because **he** is late.
$$\underbrace{\hspace{2.5cm}}_{5.34}$$

$2_\bigcirc$: The **nurse** ran because **she** is late.

(a) Coreference resolution

	A	B	$\ln \Pr[B \mid A]$
$1_\square$:	**He** is a	**doctor.**	-9.72
$1_\bigcirc$:	**She** is a	**doctor.**	-9.77
$2_\square$:	**He** is a	**nurse.**	-8.99
$2_\bigcirc$:	**She** is a	**nurse.**	-8.97

(b) Language modeling

Fig. 1. Examples of gender bias in coreference resolution and language modeling as measured by coreference scores (left) and conditional log-likelihood (right).

the method adds a copy with an *intervention* on its targeted words, replacing each with its partner, while maintaining the same, non-intervened, ground truth. The method results in a dataset of *matched pairs* with ground truth independent of the target distinction (see Fig. 1a and Fig. 1b for examples). This encourages learning algorithms to not pick up on the distinction.

Our empirical evaluation shows that CDA effectively decreases gender bias while preserving accuracy. We also explore the space of mitigation strategies with CDA, a prior approach to word embedding debiasing (WED), and their compositions. We show that CDA outperforms WED, drastically so when word embeddings are co-trained. For pre-trained embeddings, the two methods can be effectively composed. We also find that as training proceeds on the original data set with gradient descent the gender bias grows as the loss reduces, indicating that the optimization encourages bias; CDA mitigates this behavior.

In the body of this paper we present necessary background (Sect. 2), our methods (Sects. 3 and 4), their evaluation (Sect. 5), and speculate on future research (Sect. 6).

2   Background

In this section we briefly summarize requisite elements of neural coreference resolution and language modeling systems: scoring layers and loss evaluation, performance measures, and the use of word embeddings and their debiasing. The tasks and models we experiment with later in this paper and their properties are summarized in Table 1.

Coreference Resolution. The goal of a coreference resolution [5] is to group *mentions*, base text elements composed of one or more consecutive words in an input instance (usually a document), according to their semantic identity. The words in the first sentence of Fig. 1a, for example, include "the doctor" and "he". A coreference resolution system would be expected to output a grouping that places both of these mentions in the same cluster as they correspond to the same semantic identity.

Table 1. Models, their properties, and datasets evaluated.

Task/Dataset	Model	Loss via	Trainable embedding	Pre-trained embedding
Coreference resolution/CoNLL-2012 [20]	Lee et al. [14]	Coref. score		✓
	Clark and Manning [4]	Coref. clusters	✓	✓
Language modeling/Wikitext-2 [18]	Zaremba et al. [25]	Likelihood	✓	

Neural coreference resolution systems typically employ a *mention-ranking* model [5] in which a feed-forward neural network produces a coreference score assigning to every pair of mentions an indicator of their coreference likelihood. These scores are then processed by a subsequent stage that produces clusters.

The ground truth in a corpus is a set of mention clusters for each constituent document. Learning is done at the level of mention scores in the case of [14] and at the level of clusters in the case of [4]. The performance of a coreference system is evaluated in terms of the clusters it produces as compared to the ground truth clusters. As a collection of sets is a partition of the mentions in a document, partition scoring functions are employed, typically MUC, B^3 and $CEAF_{\phi 4}$ [20], which quantify both precision and recall. Then, standard evaluation practice is to report the average F1 score over the clustering accuracy metrics.

Language Modeling. A language model's task is to generalize the distribution of sentences in a given corpus. Given a sentence prefix, the model computes the likelihood for every word indicating how (un)likely it is to follow the prefix in its text distribution. This score can then be used for a variety of purposes such as auto completion. A language model is trained to minimize *cross-entropy loss*, which encourages the model to predict the right words in unseen text.

Word Embedding. Word embedding is a representation learning task for finding latent features for a vocabulary based on their contexts in a training corpus. An embedding model transforms syntactic elements (words) into real vectors capturing syntactic and semantic relationships among words.

Bolukbasi et al. [2] shows that embeddings demonstrate bias. Objectionable analogies such as "man is to woman as programmer is to homemaker" indicate that word embeddings pick up on historical biases encoded in their training corpus. Their solution modifies the embedding's parameters so that gender-neutral words no longer carry a gender component. We omit here the details of how the gender component is identified and removed. What is important, however, is that only gender-neutral words are affected by the debiasing procedure.

All of our experimental systems employ an initial embedding layer which is either initialized and fixed to some pretrained embedding, initialized then trained alongside the rest of the main NLP task, or trained without initializing. In the latter two cases, the embedding can be debiased at different stages of the training process. We investigate this choice in Sect. 5.

Related Work. Two independent work [21,27] explore gender bias in coreference resolution systems. There are differences in our goals and methods. They focus on bias in coreference resolution systems and explore a variety of such systems, including rule-based, feature-rich, and neural systems. In contrast, we study bias in a set of neural natural language processing tasks, including but not exclusively coreference resolution. This difference in goals leads to differences in the notions of bias. We define bias in terms of internal scores common to a neural networks, while both [27] and [21] evaluate bias using Winograd-schema style sentences specifically designed to stress test coreference resolutions. The independently discovered mitigation technique of [27] is closely related to ours. Further, we inspect the effect of debiasing different configurations of word embeddings with and without counterfactual data augmentation. We also empirically study how gender bias grows as training proceeds with gradient descent with and without the bias mitigation techniques.

Other related work includes the study of gender bias in other NLP applications such as neural machine translation [7,24], more recent models such as ELMo [26]. Beside gender bias, other forms of social bias in NLP have also been studied [16,17]. Some works that built on CDA method in the preprint [15] include addressing bias in other languages [29], or the analysis of counter-factual augmentation in general [12].

3   Measuring Bias

Our definition of bias loosely follows the idea of causal testing: matched pairs of individuals (instances) that differ in only a targeted concept (like gender) are evaluated by a model and the difference in outcomes is interpreted as the causal influence of the concept in the scrutinized model.

As an example, we can choose a test corpus of simple sentences relating the word "professor" to the male pronoun "he" as in sentence $1_\square$ of Fig. 1a along with the matched pair $1_\bigcirc$ that swaps in "she" in place of "he". With each element of the matched pair, we also indicate which mentions in each sentence, or context, should attain the same score. In this case, the complete matched pair is $(1_\square, (\text{professor}, \text{he}))$ and $(1_\bigcirc, (\text{professor}, \text{she}))$. We measure the difference in scores assigned to the coreference of the pronoun with the occupation across the matched pair of sentences.

We begin with the general definition and instantiate it for measuring gender bias in relation to occupations for both coreference resolution and language modeling.

Definition 1 (Score Bias). *Given a set of matched pairs D (or class of sets 𝒟) and a scoring function s, the bias of s under the concept(s) tested by D (or 𝒟), written $\mathcal{B}_s(D)$ (or $\mathcal{B}_s(\mathcal{D})$) is the expected difference in scores assigned to the matched pairs (or expected absolute bias across class members):*

$$\mathcal{B}_s(D) \overset{\text{def}}{=} \underset{(a,b)\in D}{\mathbb{E}} (s(a) - s(b)) \qquad \mathcal{B}_s(\mathcal{D}) \overset{\text{def}}{=} \underset{D \in \mathcal{D}}{\mathbb{E}} |\mathcal{B}_s(D)|$$

3.1   Occupation-Gender Bias

The principle concept we address in this paper is gender, and the biases we will focus on in the evaluation relate gender to gender-neutral occupations. To define the matched pairs to test this type of bias we employ interventions[2]: transformations of instances to their matches. Interventions are a more convenient way to reason about the concepts being tested under a set of matched pairs.

Definition 2 (Intervention Matches). *Given an instance i, corpus D, or class 𝒟, and an intervention c, the intervention matching under c is the matched pair i/c or the set of matched pairs D/c, respectively, and is defined as follows.*

$$i/c \overset{\text{def}}{=} (i, c(i)) \qquad D/c \overset{\text{def}}{=} \{i/c : i \in D\}$$

The core intervention used throughout this paper is the naive intervention g_{naive} that swaps every gendered word in its inputs with the corresponding word of the opposite gender. In Sect. 4 we define more nuanced forms of intervention for the purpose of debiasing systems.

We construct a set of sentences based on a collection of templates. In the case of coreference resolution, each sentence, or *context*, includes a placeholder for an occupation word and the male gendered pronoun "he" while the mentions to score are the occupation and the pronoun. An example of such a template is the sentence **"The [OCCUPATION] ran because <u>he</u> is late."** where the underline words indicate the mentions for scoring. The complete list can be found in the Supplemental Materials.

Definition 3 (Occupation Bias). *Given the list of templates T, we construct the matched pair set for computing gender-occupation bias of score function s for an occupation o by instantiating all of the templates with o and producing a matched pair via the naive intervention g_{naive}:*

$$D_o(T) \overset{\text{def}}{=} \{t\,[[OCCUPATION] \mapsto o] : t \in T\}\,/g_{naive}$$

To measure the aggregate occupation bias *over all occupations O we compute bias on the class 𝒟(T) where $\mathcal{D}(T) \overset{\text{def}}{=} \{D_o(T) : o \in O\}$.*

The bias measures are then simply:

$$Occupation\ Bias \overset{\text{def}}{=} \mathcal{B}_s(D_o(T))$$

$$Aggregate\ Occupation\ Bias\ (AOG) \overset{\text{def}}{=} \mathcal{B}_s(\mathcal{D}(T))$$

---

[2] Interventions as discussed in this work are automatic with no human involvement.

For language modeling the template set differs. There we assume the scoring function is the one that assigns a likelihood of a given word being the next word in some initial sentence fragment. We place the pronoun in the initial fragment thereby making sure the score is conditioned on the presence of the male or female pronoun. We are thus able to control for the frequency disparities between the pronouns in a corpus, focusing on disparities with occupations and not disparities in general occurrence. An example[3] of a test template for language modeling is the fragment **"He is a | [OCCUPATION]"** where the pipe delineates the sentence prefix from the test word. The rest can be seen in the Supplemental Materials. Since our evaluation depends on the choice of template sentences, the result might differ with different set of templates. In this paper, we specifically study gender-occupation bias, while other types of bias can be addressed by adapting the issues of interest and choices of templates.

4   Counterfactual Data Augmentation (CDA)

In the previous section we have shown how to quantify gender bias in coreference resolution systems and language models using a naive intervention, or g_{naive}. The disparities at the core of the bias definitions can be thought of as unwanted effects: the gender of the pronouns like he or she has influence on its coreference strength with an occupation word or the probability of emitting an occupation word though ideally it should not. Following the tradition of causal testing, we make use of matched pairs constructed via interventions to augment existing training datasets. By defining the interventions so as to express a particular concept such as gender, we produce datasets that encourage training algorithms to not capture that concept.

Definition 4 (Counterfactual Data Augmentation). *Given an intervention c, the dataset D of input instances (X, Y) can be c-augmented, or D/c, to produce the dataset $D \cup \{(c(x), y)\}_{(x,y)\in D}$.*

Note that the intervention above does not affect the ground truth. This highlights the core feature of the method: an unbiased model should not distinguish between matched pairs, that is, it should produce the same outcome. The intervention is another critical feature as it needs to represent a concept crisply, that is, it needs to produce matched pairs that differ *only* (or close to it) in the expression of that concept. The simplest augmentation we experiment on is the naive intervention g_{naive}, which captures the distinction between genders on gendered words. The more nuanced intervention we discuss further in this paper relaxes this distinction in the presence of some grammatical structures.

Given the use of g_{naive} in the definition of bias in Sect. 3, it would be expected that debiasing via naive augmentation completely neutralizes gender bias. However, bias is not the only concern in a coreference resolution or language modeling systems; its performance is usually the primary goal. As we evaluate performance on the original corpora, the alterations necessarily reduce performance.

---

[3] As part of template occupation substitution we also adjust the article "a".

To ensure the predictive power of models trained from augmented data, the generated sentences need to remain semantically and grammatically sound. We assume that if counterfactual sentences are generated properly, the ground truth coreference clustering labels should stay the same for the coreference resolution systems. Since language modeling is an unsupervised task, we do not need to assign labels for the counterfactual sentences.

To define our gender intervention, we employ a bidirectional dictionary of gendered word pairs such as *he:she, her:him/his* and other definitionally gendered words such as *actor:actress, queen:king*. We replace every occurrence (save for the exceptions noted below) of a gendered word in the original corpus with its dual as is the case with g_{naive}.

Flipping a gendered word when it refers to a proper noun such as *Queen Elizabeth* would result in semantically incorrect sentences. As a result, we do not flip gendered words if they are in a cluster with a proper noun. For coreference resolution, the clustering information is provided by labels in the coreference resolution dataset. Part-of-speech information, which indicates whether a word is a pronoun, is obtained through metadata within the training data.

A final caveat for generating counterfactuals is the appropriate handing of *her, he and him*. Both *he* and *him* would be flipped to *her*, while *her* should be flipped to *him* if it is an objective pronoun and to *his* if it is a possessive pronoun. This information is also obtained from part-of-speech tags.

The adjustments to the naive intervention for maintaining semantic or grammatical structures, produce the *grammatical intervention*, or g_{gra}.

5   Evaluation

In this section we evaluate CDA debiasing across three models from two NLP tasks in comparison/combination with the word embedding debiasing of [2]. For each configuration of methods we report aggregated occupation bias (marked AOB) (Definition 3) and the resulting performance measured on original test sets (without augmentation). Most of the experimentation that follow employs grammatical augmentation though we investigate the naive intervention in Sect. 5.2.

5.1   Neural Coreference Resolution

We use the English coreference resolution dataset from the CoNLL-2012 shared task [20], the benchmark dataset for the training and evaluation of coreference resolution. The training dataset contains 2408 documents with 1.3 million words. We use two state-of-art neural coreference resolution models described by [14] and [4]. We report the average F1 value of standard MUC, B^3 and $\text{CEAF}_{\phi 4}$ metrics for the original test set.

NCR Model I. The model of [14] uses pretrained word embeddings, thus all features and mention representations are learned from these pretrained embeddings. As a result we can only apply debiasing of [2] to the pretrained embedding. We evaluate bias on four configurations: no debiasing, debiased embeddings

Table 2. Comparison of 4 debiasing configurations for NCR model of [14].

Index	Debiasing configuration	Test acc. (F1)	ΔTest acc.	AOB	ΔAOB%
1.1	None	67.20[a]	–	3.00	–
1.2	CDA (g_{gra})	67.40	+0.20	1.03	−66%
1.3	WED	67.10	−0.10	2.03	−32%
1.4	CDA (g_{gra}) w/WED	67.10	−0.10	0.51	−83%

[a]Matches state-of-the-art result of [14].

(written WED), CDA only, and CDA with WED. The configurations and resulting aggregate bias measures are shown in Table 2.

In the aggregate measure, we see that the original model is biased (recall the scale of coreference scores shown in Fig. 1). Further, each of the debiasing methods reduces bias to some extent, with the largest reduction when both methods are applied. Impact on performance is negligible in all cases.

Figure 2 shows the per-occupation bias in Models 1.1 and 1.2. It aligns with the historical gender stereotypes: female-dominant occupations such as *nurse*, *therapist* and *flight attendant* have strong negative bias while male-dominant occupations such as *banker*, *engineer* and *scientist* have strong positive bias. This behaviour is reduced with the application of CDA.

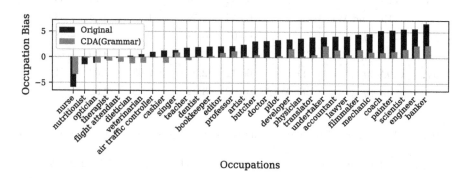

Fig. 2. Model 1.1 & 1.2: bias for occupations in original & CDA model

NCR Model II. The model of [4] has a trainable embedding layer, which is initialized with the *word2vec* embedding and updated during training. As a result, there are three ways to apply WED: we can either debias the pretrained embedding before the model is trained (written $\overleftarrow{\text{WED}}$), debias it after model training (written $\overrightarrow{\text{WED}}$), or both. We also test these configurations in conjunction with CDA. In total, we evaluate 8 configurations as in shown in Table 3.

The aggregate measurements show bias in the original model, and the general benefit of augmentation over word embedding debiasing: it has better or comparable debiasing strength while having lower impact on accuracy. In models 2.7

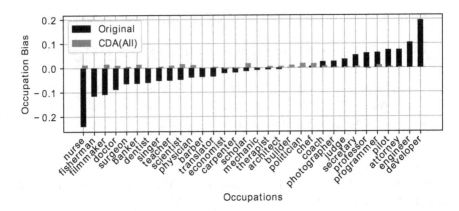

Fig. 3. Model 3.1 & 3.3: bias for occupations in original RNN language model

Table 3. Comparison of 8 debiasing configurations for NCR model of [4]. The ±AOB column is aggregate occupation bias with preserved signs in aggregation.

Index	Debiasing configuration	F1	ΔTest acc.	AOB	±AOB	ΔAOB%
2.1	None	69.10	–	2.95	2.74	–
2.2	$\overleftarrow{\text{WED}}$	68.82	−0.28	2.50	2.24	−15%
2.3	$\overrightarrow{\text{WED}}$	66.04	−3.06	0.9	0.14	−69%
2.4	$\overleftarrow{\text{WED}}$ and $\overrightarrow{\text{WED}}$	66.54	−2.56	1.38	−0.54	−53%
2.5	CDA (g_{gra})	69.02	−0.08	0.93	0.07	−68%
2.6	CDA (g_{gra}) w/$\overleftarrow{\text{WED}}$	68.5	−0.60	0.72	0.39	−75%
2.7	CDA (g_{gra}) w/$\overrightarrow{\text{WED}}$	66.12	−2.98	2.03	−2.03	−31%
2.8	CDA (g_{gra}) w/$\overleftarrow{\text{WED}}$, $\overrightarrow{\text{WED}}$	65.88	−3.22	2.89	−2.89	−2%

Table 4. Comparison of three debiasing configurations for an RNN language model.

Index	Debiasing configuration	Test perp.	ΔTest perp.	AOB	ΔAOB%
3.1	None	83.39	–	0.054	–
3.2	$\overrightarrow{\text{WED}}$	1128.15	+1044.76	0.015	−72%
3.3	CDA (g_{gra})	84.03	+0.64	0.029	−46%
3.4	CDA (g_{naive})	83.63	+0.24	0.008	−85%

and 2.8, however, we see that combining methods can have detrimental effects: the aggregate occupation bias has flipped from preferring males to preferring females as seen in the ±AOB column which preserves the sign of per-occupation bias in aggregation.

5.2   RNN Language Modeling

Although theoretically RNN language models can be either trained with a pre-trained embedding or a trainable embedding layer, the latter is more often used [22]. One reason is that since training corpus for language model are abundant, a good task-specific embedding can be learned with the rest of the model. As a result, we use the Wikitext-2 dataset [18] for language modeling and employ a simple 2-layer RNN architecture with 1500 LSTM cells and a trainable embedding layer of size 1500. As a result, word embedding can only be debiased after training. The language model is evaluated using *perplexity*, a standard measure for averaging *cross-entropy loss* on unseen text. We also test the performance impact of the naive augmentation in relation to the grammatical augmentation in this task. Figure 3 shows the per-occupation bias in Models 3.1 and 3.3. The aggregate results for the four configurations are show in Table 4.

We see that word embedding debiasing in this model has very detrimental effect on performance. The post-embedding layers here are too well-fitted to the final configuration of the embedding layer. We also see that the naive augmentation almost completely eliminates bias and surprisingly happened to incur a lower perplexity hit. We speculate that this is a small random effect due to the relatively small dataset (36,718 sentences of which about 7579 have at least one gendered word) used for this task.

5.3   Learning Bias

The results presented so far only report on the post-training outcomes. Figure 4, on the other hand, demonstrates the evolving performance and bias during training under various configurations. In general we see that for both neural coreference resolution and language model, bias (thick lines) increases as loss (thin lines) decreases. Incorporating counterfactual data augmentation greatly bounds the growth of bias (gray lines). In the case of naive augmentation, the bias is limited to almost 0 after an initial growth stage (lightest thick line, right).

5.4   Overall Results

The original model results in the tables demonstrate that bias exhibits itself in the downstream NLP tasks. This bias mirrors stereotypical gender/occupation associations as seen in Fig. 2 (black bars). Further, word debiasing alone is not sufficient for downstream tasks without undermining the predictive performance, no matter which stage of training process it is applied ($\overleftarrow{WED}$ of 2.2 preserves accuracy but does little to reduce bias while $\overrightarrow{WED}$ of 2.3 does the opposite). Comparing 2.2 ($\overleftarrow{WED}$) and 2.4 ($\overleftarrow{WED}$ and $\overrightarrow{WED}$) we can conclude that bias in word embedding removed by debiasing performed prior to training is relearned by its conclusion as otherwise the post-training debias step of 2.4 would have no effect. The debiased result of configurations 1.2, 2.5 and 3.3 show that counterfactual data augmentation alone is effective in reducing bias across all tasks while preserving the predictive power.

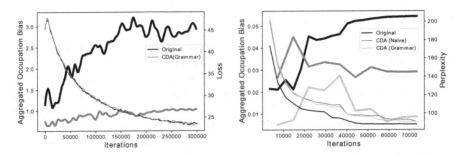

Fig. 4. Performance and aggregate occupation bias during training phases for coreference resolution with model of [14] (left) and language modeling (right).

Results combining the two methods show that CDA and pre-training word embedding debiasing provide some independent debiasing power as in 1.4 and 2.6. However, the combination of CDA and post-training debiasing has an overcorrection effect in addition to the compromise of the predictive performance as in configurations 2.7 and 2.8.

6   Future Work

A natural next step is to study bias in transformer-based language models such as BERT [6]. However, NLP tasks trained with BERT requires fine tuning on a pretrained base-model, as part of a transfer-learning pipeline. Retraining from scratch is not feasible in such pipelines as the original training process is too computationally expensive and the data for the base-model is typically unavailable. We will explore how to adapt CDA for such transfer-learning scenarios.

We are exploring other types of bias beyond gender bias, such as racial or age-related bias. One potential challenge for addressing other bias is that race and age concepts are not as easily attributable to pairs of pronouns.

Finally, we are looking into explaining why neural network models exhibit bias by studying their inner workings. Such explanations could help us encode (de)bias constraints in the training loss function to prevent bias from being introduced in the first place.

Supplemental Material

Context Template Sentences for Occupation Bias

Below is the list of the context template sentences used in our coreference resolution experiments **OCCUPATION** indicates the placement of one of occupation words listed below.

- "The [OCCUPATION] ate because <u>he</u> was hungry."
- "The [OCCUPATION] ran because <u>he</u> was late."

- "The [OCCUPATION] drove because he was late."
- "The [OCCUPATION] drunk water because he was thirsty."
- "The [OCCUPATION] slept because he was tired."
- "The [OCCUPATION] took a nap because he was tired."
- "The [OCCUPATION] cried because he was sad."
- "The [OCCUPATION] cried because he was depressed."
- "The [OCCUPATION] laughed because he was happy."
- "The [OCCUPATION] smiled because he was happy."
- "The [OCCUPATION] went home because he was tired."
- "The [OCCUPATION] stayed up because he was busy."
- "The [OCCUPATION] was absent because he was sick."
- "The [OCCUPATION] was fired because he was lazy."
- "The [OCCUPATION] was fired because he was unprofessional."
- "The [OCCUPATION] was promoted because he was hardworking."
- "The [OCCUPATION] died because he was old."
- "The [OCCUPATION] slept in because he was fired."
- "The [OCCUPATION] quitted because he was unhappy."
- "The [OCCUPATION] yelled because he was angry."

Similarly the context templates for language modeling are as below.

- "He is a | [OCCUPATION]"
- "he is a | [OCCUPATION]"
- "The man is a | [OCCUPATION]"
- "the man is a | [OCCUPATION]"

References

1. Bahdanau, D., Cho, K., Bengio, Y.: Neural machine translation by jointly learning to align and translate. arXiv preprint arXiv:1409.0473 (2014)
2. Bolukbasi, T., Chang, K.W., Zou, J.Y., Saligrama, V., Kalai, A.T.: Man is to computer programmer as woman is to homemaker? Debiasing word embeddings. In: Advances in Neural Information Processing Systems, pp. 4349–4357 (2016)
3. Caliskan, A., Bryson, J.J., Narayanan, A.: Semantics derived automatically from language corpora contain human-like biases. Science 356(6334), 183–186 (2017)
4. Clark, K., Manning, C.D.: Deep reinforcement learning for mention-ranking coreference models. arXiv preprint arXiv:1609.08667 (2016)
5. Clark, K., Manning, C.D.: Improving coreference resolution by learning entity-level distributed representations. arXiv preprint arXiv:1606.01323 (2016)
6. Devlin, J., Chang, M.W., Lee, K., Toutanova, K.: BERT: pre-training of deep bidirectional transformers for language understanding. arXiv preprint arXiv:1810.04805 (2018)
7. Font, J.E., Costa-Jussa, M.R.: Equalizing gender biases in neural machine translation with word embeddings techniques. arXiv preprint arXiv:1901.03116 (2019)
8. Graves, A.: Generating sequences with recurrent neural networks. arXiv preprint arXiv:1308.0850 (2013)

9. Graves, A., Mohamed, A.R., Hinton, G.: Speech recognition with deep recurrent neural networks. In: 2013 IEEE International Conference on Acoustics, Speech and Signal Processing (ICASSP), pp. 6645–6649. IEEE (2013)

10. Johnson, M., et al.: Google's multilingual neural machine translation system: enabling zero-shot translation. TACL **5**, 339–351 (2017). https://transacl.org/ojs/index.php/tacl/article/view/1081

11. Jozefowicz, R., Vinyals, O., Schuster, M., Shazeer, N., Wu, Y.: Exploring the limits of language modeling. arXiv preprint arXiv:1602.02410 (2016)

12. Kaushik, D., Hovy, E., Lipton, Z.C.: Learning the difference that makes a difference with counterfactually-augmented data. arXiv preprint arXiv:1909.12434 (2019)

13. Lapowsky, I.: Google autocomplete still has a hitler problem, February 2018. https://www.wired.com/story/google-autocomplete-vile-suggestions/

14. Lee, K., He, L., Lewis, M., Zettlemoyer, L.: End-to-end neural coreference resolution. arXiv preprint arXiv:1707.07045 (2017)

15. Lu, K., Mardziel, P., Wu, F., Amancharla, P., Datta, A.: Gender bias in neural natural language processing. arXiv preprint arXiv:1807.11714 (2018)

16. Manzini, T., Lim, Y.C., Tsvetkov, Y., Black, A.W.: Black is to criminal as caucasian is to police: detecting and removing multiclass bias in word embeddings. arXiv preprint arXiv:1904.04047 (2019)

17. May, C., Wang, A., Bordia, S., Bowman, S.R., Rudinger, R.: On measuring social biases in sentence encoders. arXiv preprint arXiv:1903.10561 (2019)

18. Merity, S., Xiong, C., Bradbury, J., Socher, R.: Pointer sentinel mixture models. arXiv preprint arXiv:1609.07843 (2016)

19. Mikolov, T., Chen, K., Corrado, G., Dean, J.: Efficient estimation of word representations in vector space. arXiv preprint arXiv:1301.3781 (2013)

20. Pradhan, S., Moschitti, A., Xue, N., Uryupina, O., Zhang, Y.: CoNLL-2012 shared task: modeling multilingual unrestricted coreference in ontonotes. In: Joint Conference on EMNLP and CoNLL-Shared Task, pp. 1–40. Association for Computational Linguistics (2012)

21. Rudinger, R., Naradowsky, J., Leonard, B., Van Durme, B.: Gender bias in coreference resolution. arXiv preprint arXiv:1804.09301 (2018)

22. Sundermeyer, M., Schlüter, R., Ney, H.: LSTM neural networks for language modeling. In: Thirteenth Annual Conference of the International Speech Communication Association (2012)

23. Tatman, R.: Gender and dialect bias in YouTube's automatic captions. In: Proceedings of the First ACL Workshop on Ethics in Natural Language Processing, pp. 53–59 (2017)

24. Vanmassenhove, E., Hardmeier, C., Way, A.: Getting gender right in neural machine translation. arXiv preprint arXiv:1909.05088 (2019)

25. Zaremba, W., Sutskever, I., Vinyals, O.: Recurrent neural network regularization. arXiv preprint arXiv:1409.2329 (2014)

26. Zhao, J., Wang, T., Yatskar, M., Cotterell, R., Ordonez, V., Chang, K.W.: Gender bias in contextualized word embeddings. arXiv preprint arXiv:1904.03310 (2019)

27. Zhao, J., Wang, T., Yatskar, M., Ordonez, V., Chang, K.W.: Gender bias in coreference resolution: evaluation and debiasing methods. arXiv preprint arXiv:1804.06876 (2018)

28. Zheng, J., Chapman, W.W., Crowley, R.S., Savova, G.K.: Coreference resolution: a review of general methodologies and applications in the clinical domain. J. Biomed. Inform. **44**(6), 1113–1122 (2011)

29. Zmigrod, R., Mielke, S.J., Wallach, H., Cotterell, R.: Counterfactual data augmentation for mitigating gender stereotypes in languages with rich morphology. arXiv preprint arXiv:1906.04571 (2019)

Author Index

Printed in the United States
By Bookmasters